S0-ACA-244

FLORIDA STATE
UNIVERSITY LIBRARIES

AUG 24 1998

TALLAHASSEE, FLORIDA

# Disarmament and Defence Industrial Adjustment in South Africa

**Stockholm International Peace Research Institute**

SIPRI is an independent international institute for research into problems of peace and conflict, especially those of arms control and disarmament. It was established in 1966 to commemorate Sweden's 150 years of unbroken peace.

The Institute is financed mainly by the Swedish Parliament. The staff and the Governing Board are international. The Institute also has an Advisory Committee as an international consultative body.

The Governing Board is not responsible for the views expressed in the publications of the Institute.

**Governing Board**

Professor Daniel Tarschys, Chairman  (Sweden)
Dr Oscar Arias Sánchez  (Costa Rica)
Sir Marrack Goulding  (United Kingdom)
Dr Ryukichi Imai  (Japan)
Dr Catherine Kelleher  (United States)
Dr Marjatta Rautio  (Finland)
Dr Lothar Rühl  (Germany)
Dr Abdullah Toukan  (Jordan)
The Director

**Director**

Dr Adam Daniel Rotfeld  (Poland)

**sipri**

**Stockholm International Peace Research Institute**
Frösunda,  S-169 70 Solna  Sweden
Telephone:  46 8/655 97 00
Telefax:  46 8/655 97 33
Email:  sipri@sipri.se
Internet URL:  http://www.sipri.se

# Disarmament and Defence Industrial Adjustment in South Africa

Peter Batchelor and Susan Willett

OXFORD UNIVERSITY PRESS

1998

HC
905
.Z9
D422
1998

*Oxford University Press, Great Clarendon Street, Oxford OX2 6DP*
*Oxford New York*
*Athens Auckland Bangkok Bogotá Bombay Buenos Aires*
*Calcutta Cape Town Dar es Salaam Delhi Florence Hong Kong Istanbul*
*Karachi Kuala Lumpur Madras Madrid Melbourne Mexico City*
*Nairobi Paris Singapore Taipei Tokyo Toronto Warsaw*
*and associated companies in*
*Berlin Ibadan*

*Oxford is a registered trade mark of Oxford University Press*

*Published in the United States*
*by Oxford University Press Inc., New York*

© *SIPRI 1998*

*All rights reserved. No part of this publication may be reproduced,*
*stored in a retrieval system, or transmitted, in any form or by any means,*
*without the prior permission in writing of Oxford University Press.*
*Within the UK, exceptions are allowed in respect of any fair dealing for the*
*purpose of research or private study, or criticism or review, as permitted*
*under the Copyright, Designs and Patents Act, 1988, or in the case of*
*reprographic reproduction in accordance with the terms of the licences*
*issued by the Copyright Licensing Agency. Enquiries concerning*
*reproduction outside these terms should be sent to the Rights Department,*
*Oxford University Press, at the address above.*
*Enquiries concerning reproduction in other countries should be sent to SIPRI.*

*British Library Cataloguing in Publication Data*
*Data available*

*Library of Congress Cataloging-in-Publication Data*
*Data available*

*ISBN 0–19–829413–1*

*Typeset and originated by Stockholm International Peace Research Institute*
*Printed in Great Britain on acid-free paper and bound by Biddles Ltd,*
*Guildford and King's Lynn*

# Contents

# Preface

The authors define the purpose of this book as being twofold: to provide a comprehensive empirical analysis of the wide-ranging adjustments that have taken place within South Africa's arms industry during the country's transition to democracy; and at a more general level to locate this process of defence industrial adjustment within the broader context of South Africa's changing political, strategic and economic environment. It also attempts to place the South African experience in the context of broader debates about the economic effects of military expenditure and defence industrialization and the relationship between disarmament and development in the developing countries.

Peter Batchelor and Susan Willett have extensive first-hand experience of the developing public debate and policy making in defence and security policy in the new South Africa. They examine the experience of the South African arms industry—the largest in any developing country—consequent on the process of disarmament undertaken in the country since 1989, drastic defence cuts and the transformed regional environment. They show the structural distortions introduced in the apartheid economy by the investment in a local arms industry and consider how far a 'peace dividend' has been achieved. One of their conclusions is that the reallocation of defence resources to civilian purposes as a result of defence cuts does not occur automatically. The South African experience provides a clear example of the need to construct a link between disarmament and development through explicit government policies. It also provides useful insights for other developing countries which have sizeable defence sectors and which are experiencing a process of disarmament and demilitarization.

The outcome is uncertain. Clear government policy for the arms industry has been lacking, and the opportunity to develop a comprehensive national conversion policy may already have been missed. Meanwhile the South African Government retains an interest in maintaining high military expenditure and a significant domestic arms industry and has so far defended its active policy of exporting arms, particularly to the Southern African region.

The Stockholm International Peace Research Institute is grateful to the two authors and is pleased to have the opportunity to present the results of their research.

Adam Daniel Rotfeld
Director of SIPRI
February 1998

# Acknowledgements

This book is the product of a collaborative relationship that has flourished over the past four years between the two authors. During this period of cooperation we have worked on a number of defence-related projects within South Africa which have had a strong policy orientation. Our involvement included participation in the Military Research Group/Armscor bilateral discussions on future arms trade and defence industrial strategies; consultancy work for the African National Congress (ANC) Budget Committee working on defence budget cuts in order to release resources for the Reconstruction and Development Programme; monitoring by Susan Willett of the Security Forces in the East Rand during the April 1994 elections; a joint written submission to the Cameron Commission on the costs and benefits of the arms trade in March 1995, backed up by an oral presentation by Susan Willett in June 1995; written submissions by Peter Batchelor to the consultation process on the 1996 White Paper on National Defence; and participation in the Defence Review consultative process in February 1996.

This involvement, during what can only be described as a period of momentous change, has provided us with a unique insight into the process of disarmament, demilitarization and development in the context of South Africa. We have attempted to capture these experiences in this book in order to share them with a wider audience.

As with most collaborative ventures there has been a natural division of labour. Susan Willett as the project manager conceived and successfully raised funds for the project. She has been responsible for much of the conceptual thinking behind the book, providing the overarching theoretical framework, which has attempted to draw out the links between disarmament, demilitarization and development on the one hand and the debates on security, strategic doctrine and conversion strategies on the other. Peter Batchelor has been responsible for providing the wealth of empirical data on South Africa's defence industrial base, which has captured the unique nature of the South African defence industrial adjustment process. Much of the historical content of the book is drawn from his doctoral thesis.

Although we are entirely responsible for the interpretations presented in this book, its execution would not have been possible without valuable inputs from many colleagues, friends and fellow academics with whom we have had the pleasure of working and sometimes battling in South Africa's new policy forums. We extend special thanks to Laurie Nathan, Director of the Centre for Conflict Resolution at the University of Cape Town, whose resolute pursuit of peace and disarmament is an example to us all; to Peter Vale and Paul Dunne, who provided much intellectual inspiration and valuable comments; to Thomas Ohlson as an external referee for the book; to Annette Seegers for her friendship and encouragement; to the members of the Military Research Group, particularly Gavin Cawthra, Jacky Cock, Calvin Khan, Krish Naidoo, Ian Phillips and Rocky Williams, for friendship, camaraderie and an intellectually stimulating environment for debates on defence and security in South Africa; to Paul Cornish and Jack Spence from the Royal Institute of International Affairs in London for useful comments and interesting discussions; to all those at the Bonn International Center for Conversion (BICC) for their generosity with information, support and encouragement; to the Stockholm International Peace Research Institute (SIPRI) for the help both of Elisabeth Sköns and of the editor, Eve Johansson, and for publishing the book; to Andre Buys from Armscor and Julius Kriel from the South African Defence Industry Association (SADIA) for providing us with information never before released into the public domain and for their openness and willingness to engage with us in debates on defence industrial issues despite our obvious differences; and, finally, to all our colleagues at the Centre for Conflict Resolution and Centre for Defence Studies, friends and families who have put up with us through the painful birth of this book.

However, it has only been with the generous support of a Research and Writing Grant from the John D. and Catherine T. MacArthur Foundation that we have been afforded the chance to translate our extensive experience into a book. For this opportunity we are truly grateful.

Susan Willett and Peter Batchelor
London, September 1997

... It ought to be remembered that there is nothing more difficult to take in hand, more perilous to conduct, or more uncertain in its success, than to take the lead in the introduction of a new order of things. Because the innovator has for enemies all those who have done well under old conditions, and lukewarm defenders in those who may do well under the new. This coolness arises partly from fear of the opponents, who have the laws on their side, and partly from the incredulity of men, who do not readily believe in new things until they have a long experience of them. Thus it happens that whenever those who are hostile have the opportunity to attack they do it like partisans, whilst the others defend lukewarmly ...

Niccolo Machiavelli
*The Prince*

# Acronyms

| | |
|---|---|
| ADE | Atlantis Diesel Engines |
| AEC | Atomic Energy Corporation |
| ANC | African National Congress |
| APC | Armoured personnel carrier |
| Armscor | Armaments Development and Production Corporation (after 1977 Armaments Corporation of South Africa) |
| BMATT | British Military Advisory Training Team |
| $C^3I$ | Command, control, communications and intelligence |
| CCSA | Cabinet Committee for Security Affairs |
| CODESA | Convention for a Democratic South Africa |
| COSATU | Congress of South African Trade Unions |
| CSIR | Council for Scientific and Industrial Research |
| CSSDCA | Conference on Security, Stability, Development and Co-operation in Africa |
| CWC | Chemical Weapons Convention |
| DRC | Defence Research Committee |
| DRDC | Defence Research and Development Council |
| FRELIMO | Front for the Liberation of Mozambique |
| GDP | Gross domestic product |
| GEIS | General Export Incentive Scheme |
| HEU | Highly enriched uranium |
| IAEA | International Atomic Energy Agency |
| IDC | Industrial Development Corporation |
| IFV | Infantry fighting vehicle |
| IMF | International Monetary Fund |
| IMT | Institute for Maritime Technology |
| ISDSC | Inter-State Defence and Security Committee |
| ISP | Industrial Strategy Project |
| ITSP | Industry and Technology Survival Plan |
| JMCC | Joint Military Co-ordinating Council |
| LEO | Low earth orbit |
| MK | Umkonto we Sizwe |
| MPLA | Popular Movement for the Liberation of Angola |
| MRG | Military Research Group (South Africa) |
| NCACC | National Conventional Arms Control Committee |
| NCM | National coordinating mechanism |
| NEM | Normative Economic Model |
| NGO | Non-governmental organization |
| NIDR | National Institute for Defence Research |
| NIRR | National Institute for Rocket Research |

| | |
|---|---|
| NIS | National Intelligence Service |
| NOD | Non-offensive defence |
| NPT | Non-Proliferation Treaty (1968) |
| NSMS | National Security Management System |
| OAU | Organization of African Unity |
| OECD | Organisation for Economic Co-operation and Development |
| OFS | Orange Free State |
| PWV | Pretoria–Witwatersrand–Vereeniging (region) |
| R&D | Research and development |
| RDP | Reconstruction and Development Programme |
| RENAMO | Mozambique National Resistance Movement |
| RPV | Remotely piloted vehicle |
| SADC | Southern African Development Community |
| SADCC | Southern African Development Co-ordination Conference |
| SADF | South African Defence Force |
| SADIA | South African Defence Industry Association |
| SANDF | South African National Defence Force |
| SET | Science, engineering and technology |
| SSC | State Security Council |
| TBVC | Transkei, Bophutatswana, Venda and Ciskei |
| TDP | Technology Development Programme |
| TEC | Transitional Executive Council |
| UNAVEM | United Nations Angola Verification Mission |
| UNDP | United Nations Development Programme |
| UNIDIR | United Nations Institute for Disarmament Research |
| UNITA | National Union for the Total Independence of Angola |
| UNOMOZ | United Nations Operation in Mozambique |
| UNDP | United Nations Development Programme |

## Conventions in tables

| | |
|---|---|
| $ | US dollars |
| .. | Not available or not applicable |
| – | Nil or a negligible amount |
| b. | billion |
| m. | million |
| th. | thousand |

# 1. Introduction: militarization, disarmament and demilitarization

South Africa's first democratic elections, held on 27 April 1994, symbolized a defining moment in the country's turbulent history. The 46-year old system of apartheid was finally consigned to history. Notwithstanding the momentous significance of the elections, South Africa's new democracy is a fragile entity, for although the battles for political plurality and universal suffrage have been won the war against the legacies of apartheid—namely inequality, poverty, illiteracy, crime, violence and unemployment—has only just begun. Failure to resolve these problems could seriously undermine South Africa's long-term future and peaceful transition to democracy. The challenges of the transition are compounded by the fact that the new South Africa emerged through a process of 'negotiated revolution' which means that many of the structures, institutions, and cultural and ideological attitudes of the apartheid era remain intact. Commenting on these legacies Ohlson notes,

Apartheid as a juridical system is gone but it lives on as a socio-economic structure, a security system, a lifestyle and a mental legacy. White power remains entrenched in economic and state structures. The ending of legislated apartheid is a necessary but not sufficient condition for the ending of apartheid as a system of racial domination. The contradictions, although less incompatible, remain. The difference now is that a fragile political culture advocating non-violent solutions to societal conflicts has found root and taken preliminary constitutional form.[1]

Under the apartheid system violence became the legitimate means of resolving conflict and the method by which the state maintained its power. The institutionalization of violence was secured through the militarization of the state and key sectors of civil society.[2] At the height of the apartheid era the military forces operated without transparency or accountability and often in a ruthless and clandestine manner. The state's strategic perceptions were characterized by the ideology of 'Total Strategy' which was first articulated in the 1977 White Paper on Defence.[3] The implementation of Total Strategy, which was an attempt to preserve the apartheid state against the 'total onslaught' of communist expansionism in Southern Africa, led to the militarization of South African society which began in the mid-1970s and continued throughout the 1980s.

In order to defend the South African state against internal and external threats, the apartheid government invested vast amounts of resources to develop

[1] Ohlson, T., 'South Africa: from apartheid to multi-party democracy', *SIPRI Yearbook 1995: Armaments, Disarmament and International Security* (Oxford University Press: Oxford, 1995), p. 117.
[2] For an account of the militarization of South African society in the 1980s, see Cock, J. and Nathan, L. (eds), *War and Society: The Militarisation of South Africa* (David Philip: Cape Town, 1989).
[3] South African Department of Defence, *White Paper on Defence 1977* (Government Printer: Pretoria, 1977).

a formidable war machine. This war machine was the cause of much regional instability, socio-economic dislocation, many deaths and great misery, the legacy of which visibly remains in the minefields, refugees and amputees of Angola and Mozambique. The South African military not only failed to secure the apartheid state against the Front-Line States,[4] but was also itself the cause of much internal insecurity within South Africa.

The use of state violence to defend white minority rule against real and imagined internal and external threats created a strong anti-militarist sentiment within certain sections of the African National Congress (ANC). These anti-militarist sentiments are reflected in some of the ANC's policy statements from the early 1990s:

In its dying years, apartheid unleashed a vicious wave of violence. Thousands and thousands of people have been brutally killed, maimed, and forced from their homes. Security forces have all too often failed to act to protect people, and have frequently been accused of being implicated in, and even fermenting, this violence. We are close to creating a culture of violence in which no person can feel any sense of security in their person or property. The spectre of poverty and/or violence haunts millions of our people.[5]

The growing power and influence of the military, the South African Defence Force's involvement in a number of regional conflicts, which required a guaranteed source of appropriate armaments and military equipment, and the imposition of a mandatory United Nations arms embargo in 1977[6] prompted the apartheid government to invest considerable national resources in developing a domestic arms industry with across-the-board capabilities. Despite the arms embargo, South Africa developed a strategic nuclear capability and increased its offensive conventional capabilities for military intervention in the regional context. By the late 1980s its arms industry had developed into one of the most significant sectors of the country's industrial base. Furthermore, the country had also become a major developing-country arms producer (a 'third-tier' country) and was actively engaged in the international arms trade.[7]

During its build-up from the mid-1970s onwards the arms industry became a major site for Afrikaner political and economic empowerment. The increasing allocation of scarce resources to the arms industry during the 1970s and throughout the 1980s contributed to the structural distortions and declining performance of the South African economy which started to emerge during the 1970s and 1980s. The fundamental political and strategic changes which have occurred in South Africa since 1989 have had a profound impact on the size and structure of the country's defence industrial base. The arms industry has been

---

[4] The Front-Line States were Angola, Botswana, Lesotho, Malawi, Mozambique, Swaziland, Tanzania, Zambia and Zimbabwe.

[5] African National Congress, *The Reconstruction and Development Programme: A Policy Framework* (Umanyano Publications: Johannesburg, 1994), p. 3.

[6] UN Security Council Resolution 418 (1977).

[7] Anthony, I., 'Third tier countries', ed. H. Wulf, SIPRI, *Arms Industry Limited* (Oxford University Press: Oxford, 1993), p. 365.

forced to down-size and restructure quite dramatically since the late 1980s as a result of drastic defence cuts and the formulation of new policies on defence and security matters.

In the lead-up to the elections in April 1994 the apparent ease with which negotiations took place between the apartheid government and the liberation movements over the procedures for military integration and the reformulation of defence and security priorities led to a general impression that consensus had been achieved over the military restructuring process. However, since the new ANC Government came to power certain actors and groupings have expressed their intention and commitment to abolish the legacy of militarization through the combined processes of disarmament, demilitarization and development. Such organizations include the radical churches and influential groupings within the ANC, who are intent on realizing a 'peace dividend' in order to help redress the legacies of inequality inherited from the apartheid system. These objectives are embodied in the ANC's Reconstruction and Development Programme (RDP)[8] which has the stated goal of shifting national resources away from defence towards more pressing socio-economic needs.

Tensions between the old order and the new, and within certain sections of the ANC, together with the deep mistrust the vast majority of the population has towards the military sector have meant that the formulation of new defence and security policies is being vigorously debated in the new, democratically elected parliament and other public forums. This debate has become the focus of a profound but subtle struggle for power between many influential forces within South African society. On the one hand there are the old military institutions and structures of force, such as Armscor (the Armaments Corporation of South Africa, the state-owned armaments production and procurement corporation), the South African National Defence Force (SANDF) and the defence industrialists, intent on maintaining as much of the power, influence and institutional muscle which they had under apartheid as possible. On the other hand there are the newly empowered constituencies within South African civil society, intent on deconstructing the legacy of militarization through the combined processes of demilitarization, disarmament and development.

It would be wrong, however, to present the struggle over defence resources as a simple battle between the old order and the new, for there are distinct elements within the ANC which have a vested interest in maintaining high military expenditure and a significant domestic arms industry. These include arms industry workers and former members of Umkonto we Sizwe (MK), the armed wing of the ANC, who have given voice to ambitious military objectives such as regional power projection and engagement in international peacekeeping operations.[9]

---

[8] See note 5.

[9] Two examples of senior MK officials arguing for these positions are Joe Modise, now Minister of Defence, and Col Rocky Williams, a strategic analyst within the new Defence Secretariat in Pretoria. Modise, J., 'The future a shared responsibility', *Military Technology*, Mar. 1993; and Williams, R. (Col), 'Practical challenges to South African peace operations', eds M. Shaw and J. Cilliers, *South Africa and Peace Keeping in Africa, Vol. 1* (Institute for Defence Policy: Halfway House, 1995).

After much impassioned debate between these competing forces the principles of a new defence and security policy were presented in the White Paper on National Defence, which was approved by parliament in May 1996.[10] The practical implementation of these policies was the subject of a major Defence Review, completed in the summer of 1997. The Review was concerned with the nuts and bolts of defence policy, such as force design and force structure, and is likely to have profound implications for future procurement plans and by implication for the domestic arms industry.

In early 1997 the government started the process of producing a White Paper on the Defence Industry. This will not be complete before early 1998, and the process of producing it will be characterized by a high level of debate between various stakeholders about the future of the South African arms industry.

In this book the authors attempt to capture the dynamic nature of the internal and external changes which are affecting the South African arms industry. These influences are not restricted to the defence and security field, for the general transition process within South African society involves the emergence of new norms, values and principles which are fundamentally altering the nature of South African society, forcing a re-examination of the foundations on which defence production and arms trade policies were based during the apartheid era. In particular, the new government's emphasis on human rights, both in domestic and in foreign policy, has resulted in a strong preoccupation with the ethical and moral dimensions of domestic arms production and arms sales.

Domestic politics are not the only factors exerting pressure for change within the South African arms industry. External economic and strategic circumstances are also impacting upon the arms industry. The profound changes which have taken place in the Southern African region[11] and the international security environment since 1989 have resulted in a discontinuity in defence and security thinking, which in turn is having a significant impact on the dynamics of supply and demand in the global arms market and therefore on the structure of the arms industry worldwide. South Africa, having re-entered the international community, can no longer remain isolated from these external forces. This book therefore also focuses on the contextual influences of the transformation of South African society and the impact of the changing geo-strategic environment on its defence industrial base.

While the focus of the study is primarily on the 'peculiarity' of the South African experience of defence industrial adjustment, fashioned as it is by the specific historical and cultural legacies of apartheid and by the transformation of its geo-strategic and political environment, it is hoped that it will also contribute to a better understanding of the relationship between disarmament and development. This issue lies at the heart of current debates in South Africa on the future of the country's arms industry. Apartheid created a highly structured

---

[10] South African Department of Defence, *Defence in a Democracy: White Paper on National Defence* (Government Printer: Pretoria, 1996). See chapter 6, sections I and II in this volume.

[11] Southern Africa is defined for the purposes of this book as Angola, Botswana, Lesotho, Malawi, Mozambique, Namibia, South Africa, Swaziland, Tanzania, Zambia and Zimbabwe.

**Table 1.1.** The evolution of the South African arms industry

| Period | Predominant economic and political characteristics | Political discourse/ strategic doctrine |
|---|---|---|
| Militarization 1977–89 | Militarization of the state<br>War<br>High degree of secrecy<br>Militarization of the economy<br>Build-up of the arms industry<br>Covert arms sales<br>Isolation | Realism<br>Total Strategy<br>Offensive defence |
| Structural disarmament 1989–94 | Defence cuts<br>Regional peace<br>End of conscription<br>Contraction of the arms industry<br>Arms export maximization<br>Isolation | Neo-realism<br>Offensive defence |
| Demilitarization 1994– | Regional peace<br>Regional cooperation<br>Defence cuts<br>Development<br>Force integration<br>Further contraction of the arms industry<br>Restrained arms export policy<br>Reintegration into the global community | Human-centred security<br>Non-offensive defence |

political economy in which a First-World enclave dominated a Third-World context. The arms industry formed a central element of this enclave, which could only be maintained by the system of unequal distribution and development.

# I. Methodology

In analysing the history and adjustment experiences of the South African arms industry this book follows a multi-disciplinary approach which draws on economics, politics, international relations theory and development studies. Although this involves complex layers of analysis it is designed to capture something of the institutional, ideological and cultural structures which have nurtured, sustained and transformed the arms industrial base since the mid-1970s. Using this multi-layered approach three distinct phases can be identified in the evolution of the South African arms industry: militarization, structural disarmament and demilitarization (see table 1.1). These phases can be delineated by the overarching strategic doctrine and/or political discourse of the time.

Throughout the book reference is made to the complex interplay between these different influences which impact upon the South African arms industry and shape its form and structure. By characterizing the evolution of the arms industry in these phases the authors attempt a more holistic analysis of the arms industry than is usually found in the literature on defence industrial adjustment. In particular attention is given to the security discourse of the time, which determines strategic doctrine and thus procurement patterns. This is particularly important given the break with the militarization period of the mid-1970s and 1980s and the attempt to reformulate South Africa's security agenda in the light of the new, predominantly non-military, security challenges facing the region.

In using the terms it is necessary to define what is meant by militarization, disarmament and demilitarization and their application to the South African context.

## Militarization

From the outset it is important to distinguish between the concepts of militarism and militarization.

Militarism is a widely used concept, but it has come to mean different things to different people. It may refer to anything from an aggressive foreign policy based on the inclination to go to war, to the preponderance of the military in the state, the promotion of military values in society, and the extreme case of a military dictatorship. It is also sometimes applied to countries with high military expenditure and large defence industrial sectors. However, as Smith and Smith have pointed out, there is no automatic relationship between these manifestations of militarism.[12] Some countries may have high military expenditure with little or no defence manufacturing capability, while others may have a military dictatorship but little or no propensity to go to war. To Smith and Smith the lack of a universal application of the term reduces it to a descriptive rather than an analytical term.

According to Ohlson the concept of militarism represents 'a static phenomenon, consisting of three components: (*a*) a set of values and attitudes (an ideology); (*b*) a social structure; and (*c*) a behaviour. Thus militarism in its purest form can be described as a set of attitudes and social practices which regard war and the preparation for war as a normal and desirable activity. Furthermore, militarism implies a tendency to favour or to seek violent solutions to problems and conflicts'.[13]

Militarization, on the other hand, is a dynamic social process:

an interactive process of increasing influence of the military on all levels of society. One can distinguish a military level proper, at which the increase in the means to perform military action (such as fighting wars) can be measured: an economic level in which the increased costs of the military sector can be measured: an ideological/cultural level, at which an increased importance is attributed to military values

---

[12] Smith, D. and Smith, R., *The Economics of Militarism* (Pluto Press: London, 1983), p. 11.

[13] Ohlson, T., *Embargo Non-Implemented: 25 Years of South African Arms Procurement in the Shadow of the UN Arms Embargoes*, CEA Working Papers (University of Mozambique: Maputo, 1988).

connected to the military (such as nation, security, honour, law and order) throughout society; and, finally a political level, at which increased political influence of the military is felt.[14]

Although a clear distinction would appear to exist between civil and military activities in democratic societies, this is not always the case in a highly authoritarian society where there is an encroachment of the military into normal civilian spheres of influence. Enloe has vividly described the intrusion of the military into civilian lives as a process involving both material and ideological dimensions.[15] In the material sense it encompasses the gradual encroachment of the military institution into civil society and the economy. The ideological dimension implies the extent to which such encroachments are acceptable to the population (in South Africa acceptable to the white minority) and come to be seen as common-sense solutions to civil problems.[16]

Empirical evidence suggests that the particular manifestations of militarization are contingent upon the historical and cultural context of the country in question. In the South African context Cock emphasizes militarization as a social process which involved the mobilization of resources for war.[17] This process emerged in response to the intensification of resistance to the apartheid state in the 1980s. It was conceived and executed by the South African Defence Force (SADF) which subsequently expanded and extended the power of the military into civil society.

Cock argues that the militarization of South African society operated at political, economic and ideological levels. At a political level it included the increasing political influence of the military within the state and within executive decision-making. At an economic level it included increasing levels of military expenditure, the establishment of a domestic arms industry and the emergence of a local military–industrial complex based on the convergence of interests between the state, the military and private capital in defending white minority rule. At an ideological level it included the legitimation of violence as a solution to conflict both within South Africa and in the region.[18]

## Disarmament

Simply interpreted, disarmament is associated with a reduction in arms. However, disarmament is a more complex and sometimes contradictory process which is not captured by this popular definition. There are many forms of dis-

[14] Ohlson, T., 'Strategic confrontation versus economic survival: problems of conflict resolution in Southern Africa', eds F. Deng and W. Zartman, *Conflict Resolution in Africa* (Brookings Institution: Washington, DC, 1991), p. 221. See also Skjelsbaek, K., 'Dimensions and modes of militarism', eds B. Huldt and A. Lejns, *Militarism and Militarisation*, Conference Paper no. 3 (Swedish Institute of International Affairs: Stockholm, 1983), pp. 1–21.

[15] Enloe, C., *Does Khaki Become You? The Militarisation of Women's Lives* (Pluto Press: London, 1983), p. 9.

[16] Cock, J., 'Introduction', eds Cock and Nathan (note 2), p. 2.

[17] Cock (note 16).

[18] Cock (note 16), pp. 5–12.

armament, such as a reduction in military expenditure, reduction or destruction of the stocks of certain weapon systems, a ban or limitation on the production of some types of military equipment, reduction in the numbers of military personnel, and cuts in defence research and development (R&D).[19] A disarmament process usually implies a modification of a nation's military strategies: for instance, the reduction or abandonment of certain types of weapons or defence capabilities, such as nuclear weapons or intercontinental ballistic missiles, may constitute a conscious effort to reduce a nation's offensive capabilities, thereby reducing the tensions created by an arms race. There is, however, no automatic association between the reduction in the number of arms and the reduction in offensive capabilities. The present shift from large conventional forces to smaller, high-technology rapid-deployment forces is a case in point.

At the same time, disarmament for the sake of cost savings can often have the effect of increasing the efficiency of military expenditure as defence industrial institutions try to find ways of producing the same level of output of weapon systems by eradicating waste, fraud and mismanagement.[20] They may also attempt to expand economies of scale and reduce unit costs by boosting foreign sales. Thus merely constraining the allocation of resources to the military function will not be sufficient to ensure that disarmament actually reduces the lethality or efficacy of the military. It is therefore wrong to assume, as is often done, that disarmament is synonymous with demilitarization.

It is also worth noting that there is a distinction between disarmament as defined above and structural disarmament. In defence economics this is understood as a form of disarmament which occurs in the absence of a political decision to disarm, resulting from the tendency for inflation in the defence sector to be higher than inflation in the rest of the economy and the tendencies of systems to become more expensive.

## Demilitarization

Demilitarization is a social process and the opposite of militarization. It includes disarmament in the sense of budget cuts and a reduction in arms and armed forces, but it is also a much broader concept which has political, economic and ideological dimensions. At a political level it includes attempts by the state to reassert civilian control over the armed forces. It also implies the search for a new normative and conceptual framework for developing defence and security policy at both national and regional levels. At an economic level it includes cuts in defence expenditure and the conversion of defence resources (labour, capital and technology) to civilian purposes. At an ideological level it includes attempts to transform the ideology of militarism.

---

[19] Hartley, K. and Sandler, T., *The Economics of Defence* (Cambridge University Press: Cambridge, 1995), pp. 260–83.
[20] Fontanel, J., 'The economics of disarmament', eds K. Hartley and T. Sandler, *Handbook of Defence Economics, Vol. 1* (Elsevier: Amsterdam, 1995).

The current attempts in South Africa to redefine the concept of security to make it a more inclusive, holistic notion which includes the non-military (economic, social, environmental and cultural) aspects of security, and which places human security and sustainable development at the centre of security policies, are a key aspect of the process of demilitarization which has been taking place since the ending of apartheid. In this context peace and development are becoming the key principles for determining the allocation of national resources. This new normative framework has yet to be operationalized at a national or regional level. Effective demilitarization is by no means secure while powerful vested interests from the old and new political orders remain intent on asserting South Africa's military might and as a result see the arms industry as a symbol of national prestige and military power. However, although the military and its supporting institutions may be intent on resisting the process of demilitarization, the important point here is that the advocates of demilitarization, recently empowered by the transition to democracy, have been instrumental in setting the policy agenda to which the military have had to respond. This contrasts sharply with the apartheid era, including the disarmament period, when the military were very much architects of their own destiny.

The three phases of the development of the South African defence industrial base since the mid-1970s have been used to organize the material gathered for this book. In each phase the authors attempt to capture the complex interplay of political, economic, cultural, ideological and institutional factors which have defined the evolution of the industry.

The transformation of South Africa's defence establishment is dominated by two major debates. The first concerns defence expenditure and how much defence expenditure is appropriate, or enough, in the post-apartheid era. This debate is linked to much broader debates about the economic effects of military expenditure and defence industrialization and the purported relationship between disarmament and development. The second relates to new thinking on security and the attempts to redefine South Africa's security and defence policies away from a narrow, military-centred concept of security towards a broader notion which embraces the non-military aspects of security and is focused on human security and sustainable development. These two debates are influenced by a variety of internal and external factors and have major implications for the future of South Africa's arms industry. Sections II and III consider the broad theoretical and practical dimensions of these two key debates.

## II. The economic effects of military expenditure and defence industrialization

The current public debate in South Africa on how much military expenditure is enough and the appropriate allocation of national resources is a healthy sign of democratization at work in a society that was previously characterized by authoritarianism and secrecy in military affairs.[21] At the core of the debate is the relationship between disarmament and development. This relationship has been extensively explored by academics but remains at a relatively abstract level.[22] Empirically there have been few, if any, examples of successful development ensuing from disarmament in developing countries, and it is unlikely that there will be many cases where the relationship is tested in the near future. South Africa may be an exception to this, for since 1989 military expenditure has been cut by over 50 per cent in real terms and various policy statements by the ANC-led government suggest a very real attempt to use the defence cuts as a development opportunity.[23] South Africa is also a particularly interesting case because it has a sizeable defence industrial base which has been forced to down-size as a result of the dramatic cuts in defence expenditure.

**The development crisis and redistribution**

The new government has articulated the view that South Africa's greatest security challenge lies in its development crisis.[24] For most people in South Africa feelings of insecurity arise from hunger, disease and deprivation rather than from the threat of external military attack. The government argues that 'the Reconstruction and Development Programme is the principal long-term means of promoting the well being and security of citizens and, thereby, the stability of the country. There is consequently a compelling need to reallocate state resources to the RDP'.[25] In this manner the RDP makes an explicit link between disarmament and development within the South African context.

The RDP is essentially a transition strategy, designed to address some of the major socio-economic and structural legacies of apartheid. The socio-economic legacy, particularly the huge disparities in income, expenditure and access to basic resources (health, housing and education) between blacks and whites, continues to occupy centre stage in post-apartheid economic debates. As a result most economic policy debates have tended to focus on the issues of redistribution and development.

Redistribution is defined as 'a process whereby command over goods and services is transferred from one group of persons to another without the quid pro quo of exchange'.[26] This definition has the advantage of emphasizing the dynamic nature of distribution as a process rather than as a one-off event. It also confines the issue to state policy in that it excludes the changes in command over goods and services brought about through the market system.

[21] 'Trimming back', *Financial Mail*, 21 Mar. 1997.
[22] Deger, S. and West, R. (eds), *Defence, Security and Development* (Pinter: London, 1987); and United Nations Development Programme, *Human Development Report 1992* (Oxford University Press: Oxford, 1992).
[23] *White Paper on National Defence* (note 10), p. 5. See also Ohlson (note 1), p. 140.
[24] *White Paper on National Defence* (note 10), p. 5.
[25] *White Paper on National Defence* (note 10), p. 5.
[26] Collard, D., 'Limits to redistribution: an overview', eds D. Collard *et al.*, *Income Distribution: Limits to Redistribution* (Scientechnica: Bristol, 1980).

The RDP provides the broad policy framework through which redistribution is to take place. Its key objectives include meeting basic needs through land reform, housing provision, supplying water, sanitation and electricity to homes, investment in public transport and infrastructure, developing human resources by improving the black education system and diversifying the economy away from the production of primary commodities, such as agricultural goods and minerals, into higher-value-added manufacturing.[27] In one sense the RDP can be interpreted as an old-fashioned Keynesian public-works programme designed to kick-start the South African economy into a period of growth, using the multiplier effect of government expenditure to generate employment and development, but it is also a carefully articulated social and political programme designed to redress the socio-economic legacies of apartheid. Moreover, the RDP reflects and generates high expectations from South Africa's impoverished black population. It is a central platform upon which the ANC's political performance and legitimacy are likely to be judged.

In general, redistributive policies aimed at development will only succeed if the economic and political barriers to change can be neutralized. These include the power of the 'haves' to resist redistribution and the economic constraints on change such as budgetary realities and bureaucratic inertia. Policies of redistribution can also contain significant dangers. For example, the benefits may end up in the hands of the relatively advantaged rather than in the hands of the 'targeted recipients'. Redistribution may thus produce unintended consequences such as a loss of future increases in output and goods. The power of certain state officials tasked with redistribution may also increase significantly and as a result civil liberties may be undermined. Despite these potential risks redistribution has important political benefits in a country that has a long history of racial conflict. A public perception that the state is concerned with the welfare of the most disadvantaged also helps to reduce political tensions, while a reduction in inequalities may help to legitimize the political economy, improve its moral foundations and guarantee greater internal security. In the South African context the redistribution problem is thus by its very nature both political and normative.

The irony of the challenge of redistribution in South Africa is that the only economic actors who have the resources and expertise to get the country moving again are those who benefited under the apartheid system. Thus a pragmatic compromise has been necessary in which the disadvantaged have acquiesced by allowing the advantaged in society to continue to prosper. In this compromise there are intrinsic tensions between economic and political rationality and between the political aspirations for redistribution and economic progress. There are also tensions between short-term political demands and long-term economic consequences and between the old military establishment and the new. How these various tensions are managed is critical for the success of the South African transition.

[27] *Reconstruction and Development Programme* (note 5), pp. 7–12.

In the first two years following the 1994 elections there was near-universal acceptance that the RDP was a political as well as a social necessity. The challenge for the ANC has been to fund such an ambitious programme. While foreign aid and efficiency savings from government expenditure have provided some of the resources for the RDP's ambitious housing, health and land redistribution schemes, by far the largest portion of RDP funding has come from the reallocation of government resources.[28] One of the main targets for reallocation has been the defence budget. In 1994 the ANC defence budget team were instructed to find substantial savings in the defence budget in order to release resources for the RDP.[29]

The severity of the recent defence cuts has generated a huge outcry from the military and its supporters.[30] The fierce debate about how much military expenditure is enough has often made it appear as if South Africa's security were inimical to its development.[31] At the core of this debate has been the question of the combined effects of military expenditure and defence industrialization on South Africa's economic development.

## Disarmament and development

The economic effects of military expenditure and defence industrialization have been the subject of a long-standing international debate. While there is no consensus, much of the quantitative evidence suggests that military expenditure imposes a substantial burden on economic development, particularly in developing countries.[32] The lack of consensus is rooted in the different theoretical perspectives and quantitative methods which have been used to analyse the economic aspects of military expenditure and defence industrialization.

The neo-classical approach sees the state as a 'rational actor' and examines the economic effects of military expenditure in terms of opportunity costs and market adjustments.[33] Opportunity costs can be defined as the benefits forgone by selecting one option at the expense of another or several others. The benefit that could have been derived from the alternative not chosen is the opportunity cost of the option actually selected. In a situation where resources are scarce, military expenditure may crowd out other forms of public expenditure (on education and health, for instance) and therefore have negative effects in terms of

[28] For a discussion of the debates between ANC cabinet ministers about using cuts in the defence budget to fund the RDP, see Ohlson (note 1), p. 140.

[29] Willett, S. and Batchelor, P., 'The South African defence budget', Military Research Group, Johannesburg, Mar. 1994, p. 1. Unpublished mimeograph.

[30] 'Defence in talks to ease budget cuts', *Business Day*, 29 May 1997.

[31] Williams, R. and Omar, A., 'Government must think twice before cutting defence budget', *Business Day*, 9 May 1994; and Willett, S. and Batchelor, P., 'Behind the call to arms', *Sunday Times,* 11 Sep. 1994.

[32] Dunne, P., 'The political economy of military expenditure: an introduction', *Cambridge Journal of Economics*, vol. 14, no. 4 (1990), pp. 395–404.

[33] For a discussion of the opportunity costs of defence expenditure, see Lyttkens, C. and Vedorato, C., 'Opportunity costs of defence: a comment on Dabelko and McCormick', *Journal of Peace Research*, vol. 21, no. 4 (1984), pp. 389–94.

social welfare and economic growth. Military expenditure is treated as a pure public good and the state thus attempts to balance the security benefits of increased defence expenditure with the reductions in social welfare as a result of cuts in other forms of government expenditure.[34]

The Keynesian approach focuses on the demand side in explaining the relationship between military expenditure and economic growth. It is based on the presence of a proactive, interventionist state which uses military expenditure as one aspect of state expenditure to increase output through multiplier effects in the presence of ineffective demand.[35] If aggregate demand is low relative to potential supply, then increases in military expenditure can lead to increased capacity utilization, increased profits and hence increased investment and economic growth.[36]

The Marxist approach, which has its origins in the work of Marx, Engels and Luxemburg, sees military expenditure as essential to the development and prosperity of capitalism.[37] Marxist theorists, such as Baran and Sweezy, have argued that under monopoly capitalism there is a strong and persistent tendency for aggregate economic surplus to rise and that the absorption of such a surplus is constrained by a deficiency in the effective demand for commodities.[38] Military expenditure, within such an 'under-consumptionist' framework, is therefore important because it can prevent 'realization crises' by absorbing this surplus without increasing wages, and thereby maintain capitalist profits.[39] This under-consumptionist approach is thus the only theory in which military expenditure is important both in itself and as an integral component of the theoretical analysis of the development of capitalism.[40]

The liberal approach tends to view military expenditure as economically harmful and unproductive. Central to the liberal approach is the concept of a military–industrial complex, which comprises those individuals and groups (politicians, military bureaucrats, arms salesmen and armaments producers) who are involved in the military expenditure allocation process and who have vested interests in maintaining high levels of military expenditure.[41] The presence of a military–industrial complex is seen as a distortion from a more ideal form of capitalism, in that firms in the arms industry are not subject to normal

[34] Deger, S. and Smith, R., 'Military expenditure and development: the economic linkages', *IDS Bulletin*, vol. 16, no. 4 (1985), pp. 49–54.

[35] Dunne, P., 'The economic effects of military expenditure in developing countries: a survey', eds N. Gleditsch *et al.*, *The Peace Dividend* (Elsevier: Amsterdam, 1996).

[36] Faini, R., Annez, P. and Taylor, L., 'Defence spending, economic structure and growth: evidence among countries and over time', *Economic Development and Cultural Change*, vol. 32, no. 3 (1984), pp. 487–98.

[37] Georgiou, G., 'The political economy of military expenditure', *Capital and Class*, no. 19 (1983), pp. 183–204.

[38] Baran, P. and Sweezy, P., *Monopoly Capital* (Monthly Review Press: London, 1966).

[39] Smith, R., 'Military expenditure and capitalism', *Cambridge Journal of Economics*, vol. 1, no. 1 (1977), pp. 61–76.

[40] Dunne (note 35).

[41] For a critical review of the literature on the military–industrial complex, see Fine, B., 'The military–industrial complex: an analytical assessment', *Cyprus Journal of Economics*, vol. 6, no. 1 (1993), pp. 26–51.

competitive forces and deviate from normal economic practices, such as cost minimization, maximization of profitability and process innovation. Writers such as Melman have attempted to show how military expenditure and the presence of a military–industrial complex can contribute to a decline in a country's industrial productivity and international competitiveness.[42]

The concept of the military–industrial complex is widely used to describe the complex interaction of the military, the state and the economy. However, as Fine has observed, while it is a useful descriptive concept it is not a particularly useful analytical tool for understanding the complex interaction between the military and the economy.[43] He argues that:

If [the military–industrial complex] constitutes an amalgam of linkages, such as those between the military and the economy, in association with the powers to serve corresponding interests, what makes the MIC any more significant than the education, welfare or any other lobby? The reason why the MIC might be privileged, is that subject to historical contingency, it may attain what might be thought of as a 'critical mass', a weight of linkages and agencies that render it uniquely significant to the economy.[44]

Similarly, Lovering argues that the concept of the military–industrial complex as an empirically derived entity is essentially descriptive and static. In reality the military–industrial complex operates within an economic system which is essentially dynamic and by implication must be subject to the same laws of motion.[45]

The theoretical literature also highlights the particular channels through which military expenditure is expected to influence economic growth. According to Smith and Smith these channels include resource allocation and mobilization, organization of production, socio-political structure and external relations.[46] Military expenditure can have direct opportunity costs by diverting resources from investment and welfare expenditure. It can also be used to mobilize resources to improve infrastructure and create demand in the economy. The establishment of a domestic arms industry can have a modernizing effect through linkages with other sectors of the economy. On the other hand it can lead to the development of a high-technology sector which is divorced from the rest of the civilian economy. Military expenditure can also have an impact on a country's socio-political structure by strengthening the state's coercive capacity and on its external relations by providing security.[47] However, the overall economic impact of military expenditure can only be determined empirically.

---

[42] Melman, S., *Profits without Production* (Knopf: New York, 1983); and Melman, S., *The Permanent War Economy: American Capitalism in Decline* (Simon and Schuster: New York, 1985).

[43] Fine (note 41), p. 2.

[44] Fine (note 41), p. 2.

[45] Lovering, J., 'A European military–industrial complex? Defence industry and arms conversion in the 1990s', Paper presented to the BSA Conference, University of Kent, Apr. 1992.

[46] Smith, R. and Smith, D., *Military Expenditure, Resources and Development*, Birkbeck College Discussion Paper no. 87 (Birkbeck College, University of London: London, 1980).

[47] Dunne (note 35).

The catalyst for a large number of empirical studies on the economic effects of military expenditure and defence industrialization in the context of developing countries came from Benoit's seminal paper of 1978.[48] Contrary to his own expectations he found that defence expenditure had a positive effect on the rate of growth of national product (income) in developing countries. Most of the subsequent studies on the economic effects of military expenditure in developing countries are a response to the challenges posed by Benoit's results.[49] Deger provides the most substantial analysis of the wide-ranging effects of military expenditure on growth and development and specifically addresses the economic implications of defence industrialization.[50] This study found that as military expenditure increased it influenced savings, investment and human capital formation. Using both econometric and empirical analysis, the study concludes that 'the overall effect of military expenditure is to reduce the growth rate. If we take all interdependent effects together an increase in the defence budget leads to a decrease in the growth rate through a decline in the savings rate, a fall in investment per unit of capital and a reduction in human capital formation'.[51]

On the basis of a cross-national statistical survey of developing economies, Mullins found that the propensity to develop national defence industrial capabilities depends on growth in gross domestic product (GDP), but that the accumulation of military power may undermine the prospects of economic growth over the long term.[52] The causal relationship is very clear here: industrialization, in particular heavy industrialization, permits the development of arms production if there is the political will to follow this course. However, as Ball has shown in her study of defence industrialization in developing countries, there are weak linkages between civil and arms industries; defence industrialization creates a limited number of jobs and consumes vast amounts of scarce foreign-exchange reserves.[53]

Väyrynen has pointed out that once an arms industry is established it is not a neutral agent in the economy. It often spreads new production processes and standards, encourages R&D-intensive activities and favours particular management practices; together these factors contribute to the creation of a new technological paradigm and culture which are embodied in the arms industry but operate in and influence the larger institutional, economic and social context.[54]

[48] Benoit, E., 'Growth and defence in developing countries', *Economic Development and Cultural Change*, vol. 26, no. 2 (1978), pp. 271–80.

[49] For critiques of Benoit's work, see Ball, N. and Boulding, K., 'Defence spending: burden or boon?', *War and Peace Report*, vol. 13, no. 1 (1984); and Ball, N., 'Defence and development: critique of the Benoit study', *Economic Development and Cultural Change*, vol. 31, no. 3 (Apr. 1983), pp. 506–24.

[50] Deger, S., *Military Expenditure in Third World Countries: The Economic Effects* (Routledge and Keegan Paul: London, 1986).

[51] Deger (note 50), pp. 244–45.

[52] Mullins, A. F., *Born Arming: Development and Military Power in New States* (Stanford University Press: New York, 1987), pp. 106–107.

[53] Ball, N., *Security and Economy in the Third World* (Princeton University Press: Princeton, N.J., 1988).

[54] Väyrynen, R., *Military Industrialisation and Economic Development: Theory and Historical Case Studies* (UNIDIR: Geneva, 1992), p. 102.

Military expenditure has not only been examined in terms of its relationship to aggregate levels of investment and growth, but also in the way in which it affects the structure of investment and technological innovation. Through procurement expenditure military expenditure has a direct effect on specific industrial sectors and on technological performance.[55] Most military equipment is procured from the high-technology sectors of aerospace and electronics. These sectors devote a substantial share of their output to defence. The attributes and organization of the military in the late 20th century have come to be defined as much by technological characteristics as by historical legacy, resource allocation and institutional structures. In the latter days of the cold war the NATO allies responded to the Soviet threat with 'force multipliers' requiring a huge mobilization of scarce technological resources—a view that was reinforced by the experience of the Persian Gulf War of 1991, which was ostensibly won through technological superiority. The use of industrial and technological resources for the enhancement of warfare has, however, been accompanied by the militarization of technology. Paukert and Richards argue that the subsequent emphasis on profit maximization from defence contracts, at the expense of investment in cost-minimizing investment in civil technologies, has a negative effect on overall economic activity in high-military-expenditure economies. The complex relationship between the military, technology and economics has become the focus of much academic research in the post-cold war era.[56]

Another analytical approach, particularly in the context of the severe resource constraints that characterize so many developing economies, has been to analyse the opportunity costs of military expenditure. This approach more often than not concentrates on the effects on industrial competitiveness and efficiency, private-sector crowding-out and trade-offs with social welfare. There are two ways of looking at opportunity costs. The traditional approach uses opportunity costs as a marginalist concept: the opportunity cost of defence is, for example, the number of units of education or social welfare sacrificed in order to increase defence output by one unit. The other way of analysing opportunity costs is to examine the total amount of resources devoted to an activity. In a well-functioning economy the opportunity cost of an activity will quite simply be the expenditure for that purpose. With the latter approach the larger the sum spent on defence the higher the opportunity costs of defence.

During the 1970s and 1980s the United Nations produced a number of reports which explored the opportunity costs argument in terms of resources that might otherwise have been used to meet social needs.[57] Comparison of the 1977 and

[55] Paukert, I. and Richards, R., 'Introduction', eds I. Paukert and R. Richards, *Defence Expenditure, Industrial Conversion and Local Employment* (International Labour Organization: Geneva, 1991), p. 4.
[56] Samuels, R., *Rich Nation, Strong Army: National Security and the Technological Transformation of Japan* (Cornell University Press: Ithaca, N.Y., 1994); Sandholtz, W. *et al.*, *The Highest Stakes: The Economic Foundations of the Next Security System* (Oxford University Press: New York, 1992); and Alic, J. *et al.* (eds), *Beyond Spinoff* (Harvard Business School Press: Boston, Mass., 1992).
[57] United Nations, *Reduction of Military Budgets: Measurement and International Reporting of Military Expenditures: Report prepared by the Group of Experts on the Reduction of Military Budgets,* Report of the Secretary General, UN document A/31/222/Rev.1, Sales no. E.77.I.6 (1977); United Nations, *Economic and Social Consequences of the Arms Race and Disarmament,* Report to the Secretary-

1989 reports presents interesting evidence on the different effects of military expenditure during different cyclical phases of the global economy. The 1977 report examined the effect of military expenditure during a period of economic growth, while the 1988 report discussed it in the context of the 'stagflation' of the late 1970s and early 1980s. During the boom of the early 1970s the factors of production were fully utilized, and in this context military expenditure had an observable effect on patterns of investment and consumption. In contrast, during a recession unemployed resources were utilized by the military without directly affecting those required by the civil sector except in some bottleneck situations. This suggests that the impact of military expenditure on growth depends on the extent of excess capacity in the economy. This has led some analysts to argue that military expenditure is likely to stimulate inadequate demand when there is idle productive capacity. However, as the 1988 report points out, increasing expenditure on arms is not an efficient way of combating recession as it draws heavily on government resources that would otherwise be allocated to health, education and welfare, with resulting social consequences.[58]

The United Nations Development Programme (UNDP) *Human Development Report 1994* goes so far as to suggest that 'arms expenditure [in developing countries] undermines human security by eating up precious resources that could have been used for human development'.[59] At times of structural adjustment much deeper cuts have been made in social expenditure than in military expenditure in many developing countries. The loss of potential spending on human development has been enormous. Some of the poorest African countries spend more on defence than they do on their populations' education and health. In connection with this issue of distorted priorities, the UNDP has expressed concern over South Africa's stated intention to pursue a policy of maximizing arms exports in Sub-Saharan Africa since an increasing proportion of regional GDP has been devoted to arms purchases in the past few years.[60] The UNDP maintains that 'countries spending very little on defence and much more on human development have been more successful at defending their national sovereignty than those spending heavily on arms. Compare the relatively peaceful experiences of Botswana, Costa Rica and Mauritius with the conflicts

---

General, UN document A/32/88/Rev.1, Sales no. E.78.IX.1 (1978); United Nations Educational, Scientific and Cultural Organization, *Social and Economic Consequences of the Arms Race and Disarmament: Review of Research Trends and Annotated Bibliography*, Reports and Papers in the Social Sciences, no. 39 (UNESCO: Paris, 1978); United Nations, Economic and Social Consequences of the Arms Race and of Military Expenditures, UN document A/37/386 (1983); United Nations, Study on the Economics and Social Consequences of the Arms Race and Military Expenditures, UN document A/43/368 (1988); and United Nations, Study on the Economic and Social Consequences of the Arms Race and Military Expenditures, UN document A/43/368 (1989). The lack of resources to meet basic human needs is well documented by the UN. The evidence includes poverty, high infant mortality, and lack of adequate housing, health care, clean water, sanitation, education and so on.

[58] Study on the Economics and Social Consequences of the Arms Race and Military Expenditures (note 57).

[59] United Nations Development Programme, *Human Development Report 1994* (Oxford University Press: Oxford, 1994), p. 50.

[60] The proportion of GDP devoted to military purchases in sub-Saharan Africa increased from 0.7% in 1960 to 3% in 1991. *Human Development Report 1994* (note 59), p. 50.

afflicting Iraq, Myanmar and Somalia'.[61] To this list should be added apartheid South Africa.

From the review of the literature it is obvious that simple international comparisons of military expenditure and economic growth lead to directly opposite results depending on the time-frame of the studies. This creates the need to distinguish between the short- and long-term economic effects of military expenditure on growth and development.[62] It is also apparent that the economic effects of military expenditure are extremely complex and historically contingent.

Dunne has argued that the role of a domestic arms industry in economic development is complex and depends on the interaction of economic, political and social variables within a specific historical context.[63] It is therefore necessary to undertake detailed empirical analysis at the abstract aggregate level and at specific disaggregated levels within a given economy. This involves analysing the defence industrial sector for the effects of military expenditure on industrial structure, technology, firm behaviour and labour markets. At the firm level the influence of defence procurement on corporate strategies and structure must be assessed. In relation to technology, defence expenditure needs to be analysed in terms of its effect on innovation, in both the military and the civilian sectors. In terms of employment, military expenditure needs to be assessed for its impact on skills and the structure of the labour market.

When analysing the effect of changes in military expenditure on the industrial and technology base, it is important to recognize that an understanding of the nature and impacts of military expenditure is a necessary starting-point but not sufficient to capture the complete process of defence industrial adjustment. Modifications in military expenditure produce changes in the economy which may require structural adjustments and which are contingent both on the specific situation and on the general policy environment.

Finally, a major limitation of the existing literature on the economic effects of military expenditure is the fact that demand for the arms industry's products is governed by political and geo-strategic factors, so that approaches that concentrate primarily on the economic aspects of militarism are inevitably restrictive, particularly at a time of major socio-political change, as exemplified by the current Southern African security environment.[64] This suggests the need to adopt a systemic as well as historically contingent analysis of military expenditure and defence industrialization within the South African context.

Finally, a major preoccupation of the current debate on South African military expenditure has been the question how much military expenditure is enough. This question is linked to: (*a*) available resources, given the change in national spending priorities; (*b*) monetary and fiscal pressures; (*c*) external donor pressure to keep military expenditure down; (*d*) the size and structure of

[61] *Human Development Report 1994* (note 59).
[62] Edelstein, M., 'What price the cold war? Military spending and private investment in the US, 1946–79', *Cambridge Journal of Economics*, vol. 14, no. 4 (1990), pp. 421–38.
[63] Dunne (note 32), p. 402.
[64] An exception is to be found in Sandholtz *et al.* (note 56).

the post-apartheid armed forces; (*e*) procurement policy; (*f*) the extent of national subsidization of the arms industry and arms trade; and (*g*) the role and cost of the military in national and regional security provision, including any role the South African military might play in international peacekeeping forces. Since the elections these issues have been widely aired by the South African media, hotly debated in parliament and the subject of some academic attention. Despite the contentious nature of this debate there appears to be a national consensus that military expenditure should stabilize at around 2 per cent of GDP. By imposing this limit the South African state faces the daunting task of having to restructure its forces and adapt its defence industrial base to a greatly reduced resource base. Inevitably those most affected by the cuts in military expenditure maintain that the resources allocated to the defence budget are not enough to sustain an efficient national defence force.

## III. A new security agenda

A conspicuous feature of the post-apartheid era has been the process of demilitarization which has been occurring at the political, economic and social levels. At a political level, there has been a conscious effort to assert civilian control over the military and to abolish its power and influence within state decision making. At an economic level there have been dramatic cuts in the defence budget and significant attempts to reallocate the savings from the defence cuts to the RDP in order to give effect to the purported link between disarmament and development.

The debates which have accompanied the recent defence cuts are a struggle not only over economic redistribution but also over institutional power, for without a redistribution of power away from the military real development is unlikely to take place. In this sense constructing the relationship between disarmament and development represents a reversal of cumulative and deeply rooted social processes that have become embedded in the defence industrial base, force structures, strategic doctrines and defence budgets. Defence cuts alone cannot guarantee a transformation of this military legacy. If disarmament is to bring about sustainable peace and development in the country and the region as a whole it must be accompanied by an active process of demilitarization of the security agenda involving the deconstruction of the ideological, cultural and institutional manifestations of the apartheid war machine.

The ending of apartheid and South Africa's reintegration into the international community have led to a major rethinking of its foreign and security policies.[65] The debates which have accompanied this re-evaluation have profound implications for the South African defence industrial base. At the heart of the new government's foreign policy concerns has been the desire to throw off

---

[65] Booth, K. and Vale, P., 'Security in Southern Africa: after apartheid, beyond realism', *International Affairs*, vol. 71, no. 2 (1995), pp. 285–304. For a discussion of the debate on security, see van Aardt, M., 'Doing battle with security: a Southern African approach', *South African Journal of International Affairs*, vol. 3, no. 2 (summer 1996), pp. 13–28.

the stigma of a pariah state—a reputation indirectly linked to the operations of the arms industry during the apartheid era—in order to become a respected member of the international community.

The return from isolation at a major historical juncture (the end of the cold war) has meant that many of the international debates on the post-cold war security environment have a far greater resonance in South Africa, because of the general policy vacuum, than they might have for more established members of the international community. Thus a distinctive feature of the South African debate about new thinking on security is the number of new security concepts which are being tried and tested there while they remain at a high level of theoretical debate within the academic and policy circles of the North. It is not uncommon to find senior military officers and key policy advisers referring to post-modernist armies, 'legitimization crises' and other concepts associated with 'new security thinking'. This phenomenon is partly attributable to the fact that the political upheaval caused by the end of apartheid has released a wave of creative thinking and policy experimentation.

For South Africa the simultaneous ending of apartheid and the cold war represents a major break with the past, heralding a more benign and cooperative regional security environment. Southern Africa had long been the focus of superpower intervention and ideological rivalry. It has recently emerged from a period characterized by high levels of intra- and interstate conflict into a period of reconciliation, peace, democratization and demilitarization. This new era has already witnessed significant regional efforts to promote sustainable peace and development in Southern Africa.[66]

Peace is evident in the growing evidence of regional disarmament and lower levels of military expenditure. Regional military expenditure declined by about one-third between 1993 and 1996,[67] reflecting a broader global trend. The international demand for weapon systems declined by over 70 per cent at an average rate of 12 per cent over the period 1984–94.[68] These global and regional indices of disarmament have given rise to great expectations of a peace dividend which could be used to rectify the inequities of the international system. The UN has attempted to support the international process of disarmament by passing resolutions which guaranteed a country's territorial integrity and setting up mechanisms (such as the UN Register of Conventional Arms) to control and reduce arms transfers. Other multilateral institutions such as the World Bank and the International Monetary Fund (IMF) have started to consider the possibility of tying financial aid to reductions in military expenditure in an attempt to reduce the risk of war among developing nations while at the same time accelerating economic and social development. These international pressures for disarma-

[66] Mills, G., 'South Africa and Africa: regional integration and security co-operation', *African Security Review*, vol. 4, no. 2 (1995).
[67] International Institute for Strategic Studies, *The Military Balance 1994–1995* (Oxford University Press: Oxford, 1995); and *The Military Balance 1996–1997* (Oxford University Press: Oxford, 1997).
[68] US Arms Control and Disarmament Agency, *World Military Expenditures and Arms Transfers* (US Government Printing Office: Washington, DC, 1995), p. 9.

ment and lower levels of military expenditure started to gain support in Southern Africa after 1989.

Given these trends, a new confidence about the prospects for world peace has found expression in the Southern African context. This new-found confidence is based on three basic principles. The first is that most leaders recognize that war does not pay and that modern states will do almost anything to avoid general war.[69] The second is that economic interdependence renders war unprofitable.[70] If a country's main markets are likely to be destroyed it is unlikely to initiate wars for reasons of vested interests. The third argument is that democracies rarely if ever go to war with one another.[71]

On the basis of these premises new notions of security have emerged which emphasize 'common' and 'comprehensive' security. Their proponents argue that military-centred notions of national security had become fundamentally flawed in a highly interdependent world facing multiple security threats that are not amenable to traditional statist solutions.[72] Common security assumes that there are global dangers which threaten the entire system and which cannot be solved by boundary protection: by emphasizing common dangers, it bases its appeal for cooperative behaviour not on altruism but on the larger sense of collective self-interest. Efforts to promote 'collective' security have been pursued in various multilateral organizations such as the Organization for Security and Co-operation in Europe (OSCE), the ASEAN (Association of South-East Asian Nations) Regional Forum (ARF), and indeed in Southern Africa the Southern African Development Community (SADC).[73]

In this new geo-strategic environment many of the traditional assumptions which formerly guided South Africa's defence and security policies have been found wanting. The notions of the sovereignty of states and national interest, the traditional principles around which South Africa's security policies were formulated, are increasingly recognized as having limited applicability in the post-apartheid world.

According to Hoffman the system of sovereign states is being undermined by two fundamental changes in the global system which are outside sovereign states' control.[74] The first is that the current changes brought about by interdependence and globalization deprive the state of much of its operational role

[69] Mansfield, E., *Power, Trade and War* (Princeton University Press: Princeton, N.J., 1992).

[70] Gilpin, R., 'The economic dimensions of international security', ed. H. Bienen, *Power, Economics and Security: The United States and Japan in Focus* (Westview Press: Boulder, Colo., 1992).

[71] Rummel, J., 'Libertarian propositions on violence within and between nations', *Journal of Conflict Resolution*, vol. 29, no. 3 (Sep. 1985); and Cashman, G., *What Causes War? An Introduction to Theories of International Conflict* (Lexington Books: New York, 1992).

[72] Bazan, B., *People, States and Fear: An Agenda for International Security Studies in the Post-Cold War Era* (Pinter: London, 1990).

[73] The SADC was founded by independent states of Southern Africa (excluding Rhodesia, South West Africa and South Africa) in 1980 as the Southern African Development Co-ordination Conference (SADCC). After the Windhoek Declaration of 1992 it was renamed the Southern African Development Community. Its current members include Angola, Botswana, Lesotho, Malawi, Mauritius, Mozambique, Namibia, South Africa, Swaziland, Tanzania, Zambia and Zimbabwe.

[74] Hoffman, S., 'The politics and ethics of military intervention', *Survival*, vol. 37, no. 4 (winter 1996), p. 31.

and transfer many of its previous functions to a largely privately controlled world economy that is beyond national regulation and under very limited inter-state control. Trends towards globalization reflected in economic interdependence (in trade, investment and financial services) and global telecommunications and media networks have all contributed to undermining the notions of sovereignty and national interest. The second major change is the normative revolution that is corroding the substance of sovereignty and restrains the rights, which originate in sovereignty, of states to be free of interference in their internal affairs. This change is linked to the human-rights element of the present shift in security thinking. Human-rights concerns have become increasingly linked to arms trade controls, as all too often arms transfers have been used by states against their own populations. This issue has particular resonance in South Africa because of the regular use of domestically produced armaments against South African citizens and the destructive legacy of South African weapons within the region, in particular landmines and small arms.

A third change might be added to Hoffman's list, namely the changing nature of power. To all intents and purposes military power appears less appropriate than it once was in the mutable dynamics of international relations.[75] Halliday contends that economic power is now the main form of international power, which need not necessarily lead to increasing political or military competition as traditionalists would have us believe.[76] Indeed, a combination of shared prosperity and good diplomacy ensures a more cooperative international climate; hence the shift towards greater economic, political and military cooperation. The shift in balance from military towards economic power has important policy implications for those countries with large arms industries, because the strong economies of the late 20th century, such as Germany and Japan, are those that have prioritized the development of civil high-technology industries rather than 'feather-bedding' their defence sectors.[77]

Ultimately the complexities and uncertainties of the contemporary international environment which South Africa officially re-entered in May 1994 complicate and impinge upon domestic choices about security and the future of its defence industrial base. Arguably South Africa has little choice but to align itself with its neighbours if it wishes to address the new security threats of environmental degradation, migration, refugees, water shortages and AIDS. Global or regional problems such as these cannot be effectively addressed by states acting as independent actors, nor can they be solved by military institutions alone. New mechanisms need to be established to meet South Africa's new security challenges.

The question how the South African arms industry is reorienting itself in the context of these multi-dimensional pressures is the subject of this book. To do

---

[75] See, e.g., Nye, J., 'Soft power', *Foreign Policy*, no. 80 (fall 1990).

[76] Halliday, F., 'The end of the cold war and international relations: some analytical and theoretical conclusions', eds K. Booth and S. Smith, *International Relations Theory Today* (Polity Press: Oxford, 1995), p. 57.

[77] Samuels (note 56).

justice to it requires a considerable amount of historical background. In order to provide a context for understanding the current state of South Africa's arms industry, chapter 2 provides an overview of the development of the South African arms industry during the period of militarization between 1975 and 1989. Chapter 3 deals with the changing strategic, political and economic environment which South Africa confronted between 1989 and 1994, a period which was marked by the ending of apartheid and witnessed a process of disarmament reflected in major cuts in military expenditure. Chapter 4 examines the defence cuts and disarmament measures implemented between 1989 and 1994 and the impact of the cuts on the domestic arms industry and the national economy. Chapter 5 analyses the adjustment experiences of public- and private-sector defence firms during this transition period. Chapter 6 examines the demilitarization of South African society in the post-apartheid era through the transformation of the security establishment, the attempts to reformulate its security agenda by placing non-military threats to security and development at the heart of the country's security policies, and the implications for strategic doctrine. It also examines the embryonic forms of regional security cooperation that are emerging, analyses how the arms industry has attempted to realign itself in the context of these national and regional policy developments, and looks at the equipment needs of the armed forces in the light of the 'new' concept of security.

Chapter 7 analyses the political and institutional changes that have been implemented within the defence establishment in order to rebalance civil–military relations in favour of civilian political authority and examines the impact of these changes on the domestic arms industry. It considers the adjustment experiences of defence firms in the post-April 1994 period, particularly the internationalization of the industry through joint ventures and strategic alliances, and the consequences of the failure of the new government to develop a coherent defence industrial policy, and highlights the erosion of shared interests which previously bound together the apartheid-era military–industrial complex. In the absence of government policy on the arms industry, chapter 8 describes the defence conversion debate in South Africa and the ways in which this debate is linked to the 'redistributive' ideals of the Reconstruction and Development Programme. The final chapter provides a summary of the findings of the book and some concluding comments on the outstanding policy dilemmas facing the new government with respect to defence industrial policy.

# 2. Militarization and the development of the arms industry

## I. Introduction

During the 1970s and 1980s many commentators argued that South Africa was experiencing a form of 'total war' which involved the escalating militarization of society and the increasing use of state violence in defence of apartheid.[1] This process of militarization 'operates at political, economic and ideological levels, affects every area of society, and impinges on the lives of all South Africans'.[2] At an economic level it involved increasing levels of military expenditure, the development of a domestic arms industry and growing institutional links between the state, the military and private industry.[3] The increasing militarization of the economy was reflected in the country's military expenditure and arms production. The indicators in table 2.1 show sustained real increases between 1961 and 1989. The most significant increases took place after 1975 and coincided with the adoption of the Total Strategy doctrine.

The dramatic increases in military expenditure from the mid-1970s were accompanied by significant changes in the structure of South Africa's defence budget (see table 2.2). The share of procurement (equipment) rose dramatically from 11 per cent in 1961 to nearly 70 per cent in 1976, the year before the imposition of the second UN arms embargo. It then declined and remained between 40 per cent and 50 per cent before increasing to nearly 60 per cent in 1989. The large share of the defence budget devoted to procurement expenditure after 1970 and throughout the 1980s was related to the development and expansion of the domestic arms industry which took place during this period.

## II. The origins and development of the arms industry

The development and expansion of South Africa's arms industry were influenced by a number of strategic, political and economic factors. Strategic factors, such as the imposition of UN arms embargoes in 1963 and 1977, the presence of growing external 'threats' to apartheid, the increasing hostility of the international community and the imposition of embargoes and sanctions, were its primary determinants. Landgren suggests that South Africa is not typical among developing countries producing armaments in that it remains the only country which was forced to develop a domestic arms industry through

[1] Cock, J. and Nathan, L. (eds), *War and Society: The Militarisation of South Africa* (David Philip: Cape Town, 1989).
[2] Cock and Nathan (note 1), p. xiii.
[3] Cock, J., 'Introduction', eds Cock and Nathan (note 1), p. 5.

**Table 2.1.** Military expenditure and arms production, 1961–89

|  | 1961 | 1970 | 1975 | 1980 | 1985 | 1989 |
|---|---|---|---|---|---|---|
| Military expenditure (m. rand, constant 1985 prices) | 617 | 1 627 | 3 527 | 3 546 | 4 274 | 5 711 |
| Military expenditure per capita (m. rand, constant 1985 prices) | 35 | 72 | 138 | 125 | 135 | 154 |
| Military expenditure as share of GDP (%) | 1.3 | 2.1 | 3.6 | 3.1 | 3.5 | 4.1 |
| Military expenditure as share of government expenditure (%) | 7.7 | 9.8 | 14.2 | 13.2 | 12.2 | 13.0 |
| Domestic arms production (m. rand, constant 1985 prices) | 26 | 184 | 567 | 1 705 | 1 386 | 2 056 |

*Sources:* South African Reserve Bank, *Quarterly Bulletin* (South African Reserve Bank: Pretoria, various issues); and private communications with Armscor officials (domestic arms production).

import substitution over a long period of time, while all the other major developing-country producers were able to continue to import armaments.[4]

Political factors such as the implementation of discriminatory and repressive apartheid policies after 1961, which led to increasing black resistance and civil unrest, forced the state to develop a domestic military production capability in order to supply the security forces (both the police and later the armed forces) with the means to maintain minority rule.[5] The escalating militarization of South African society during the 1970s and 1980s, as reflected in rises in military expenditure and the growing political power and influence of the military establishment, was also an important determinant of domestic arms production. A significant feature of this process of militarization at an economic level was the development of shared interests between the state, the military and the private sector.[6] The private sector's interests in maintaining political and economic stability in the face of increasing domestic unrest required the maintenance of the state's military capabilities, which in turn led to the growing involvement of the private sector in domestic arms production.[7]

South Africa's direct and indirect military involvement in a number of regional conflicts during the 1970s and 1980s, such as those in Angola and Namibia, required a guaranteed and continuous supply of armaments and equipment geared to local conditions, and thus also provided the impetus for the development of a domestic arms production capability.

[4] Landgren, S., SIPRI, *Embargo Disimplemented* (Oxford University Press: Oxford, 1989), p. 12. Between 1945 and 1990 the UN Security Council imposed arms embargoes on only 2 countries: South Africa and Southern Rhodesia. The embargo against Southern Rhodesia was lifted in 1980 with the independence of Zimbabwe.
[5] Cock (note 3), pp. 1–13.
[6] Philip, K., 'The private sector and the security establishment', eds Cock and Nathan (note 1), pp. 202–16.
[7] Simpson, G., 'The politics and economics of the armaments industry in South Africa', eds Cock and Nathan (note 1), pp. 217–31.

**Table 2.2.** The distribution of military expenditure by function, 1961–89
Figures are percentages of total military expenditure.

|  | 1961 | 1970 | 1975 | 1980 | 1985 | 1989 |
|---|---|---|---|---|---|---|
| Personnel | 33.8 | 28.4 | 16.9 | 23.5 | 25.1 | 18.8 |
| Operating | 54.5 | 25.9 | 17.1 | 26.6 | 30.4 | 22.6 |
| Procurement[a] | 11.7 | 45.7 | 66.0 | 49.9 | 44.5 | 58.5 |
| **Total** | **100.0** | **100.0** | **100.0** | **100.0** | **100.0** | **100.0** |

[a] All procurement expenditure is funded from the Special Defence Account.

*Source:* South African Department of State Expenditure, *Printed Estimates of Expenditure, State Revenue Account* (Department of State Expenditure: Pretoria, various years).

The development of an industry and the achievement of self-sufficiency in arms production were also linked to the interests of Afrikaner nationalism and the emergence of Afrikaner capital.[8] While self-sufficiency in strategic industries, such as armaments, was seen as a strategic necessity in the context of embargoes, it was also seen as a powerful expression of South Africa's independence from foreign interference. The establishment of state enterprises was regarded as an integral part of a broader policy of providing jobs for (white) Afrikaners and promoting the development of Afrikaner industrial capital.

The economic boom of the 1960s and early 1970s provided the South African state with the much-needed resources to fund an explicit industrial strategy of investment in strategic industries such as oil and armaments. By 1977 South Africa's industrial sector was sufficiently large and diversified to support the establishment of a large domestic arms industry. State investment in strategic industries was also regarded by the government as a vital part of a broader policy of import substitution which was intended to promote self-sufficiency and industrialization, save on foreign exchange and contribute to South Africa's technological development. Investment in arms production was seen as a way of stimulating inter-industrial linkages and contributing to the diversification and expansion of South Africa's industrial base.

The expansion of the domestic arms industry also contributed to the development of closer links between the state and capital. With the deteriorating performance of the economy after 1974 and South Africa's growing international isolation, the state and capital moved closer together in an attempt to consolidate and protect their own interests. In this context the private sector viewed involvement in arms production as an 'easy' way of utilizing surplus capacities and maintaining profitability. The fact that the strategic concerns of government increasingly coincided with the short-term economic interests of private capital was an important determinant. By the end of the 1980s South Africa had transformed itself from an arms importer with a limited domestic arms production capability into a relatively self-sufficient arms producer. During the 1980s it

[8] Lipton, M., *Capitalism and Apartheid* (David Philip: Cape Town, 1986).

**Table 2.3.** Military industrialization: stages of development

| Stage | Arms production: capability and independence |
|---|---|
| 1 | Maintenance of equipment and weapon systems |
| 2 | Overhaul, refurbishment and rudimentary modification of equipment and weapon systems |
| 3 | Assembly of imported components and some simple licensed production |
| 4 | Local production of components and raw materials |
| 5 | Final assembly of simple weapon systems and local production of components and sub-assemblies |
| 6 | Co-production or complete licensed production of some weapon systems |
| 7 | Limited R&D improvements or modifications to local licence-produced arms |
| 8 | Limited independent production of simple weapons |
| 9 | Independent R&D and production of simple weapon systems |
| 10 | Independent R&D and production of advanced weapon systems with foreign high-technology components |
| 11 | Self-sufficiency: completely independent R&D and production of advanced weapon systems |

*Source:* Krause, K., *Arms and the State: Patterns of Military Production and Trade* (Cambridge University Press: Cambridge, 1992).

emerged as one of the developing world's major arms producers and a leading exporter of armaments and military equipment.[9] Entry into the international arms market during the early 1980s, coinciding with the government's attempts to promote export-led economic growth in manufacturing, also contributed to the ongoing development and expansion of the domestic arms industry.

## Military industrialization: stages of development

The literature on military industrialization in developing countries identifies a number of common stages of development which countries progress through as they become more self-sufficient and independent in armaments production.[10] Krause identifies 11 stages of development in a country's military industrialization process (see table 2.3).

The development of South Africa's arms industry between 1961 and 1989 can be divided into three stages: (a) 1961–68, up to stage 5, (b) 1968–77, up to stages 7–8; and (c) 1977–89, to stages 9–10. In each it was influenced and constrained by political and strategic developments and by the performance of the South African economy, particularly that of the manufacturing sector.

[9] For details of South Africa's arms industry in relation to those of other developing countries, see Brzoska, M., 'South Africa: evading the embargo', eds M. Brzoska and T. Ohlson, SIPRI, *Arms Production in the Third World* (Taylor & Francis: London, 1986); and Brauer, J., 'Arms production in developing nations: the relation to industrial structure, industrial diversification and human capital formation', *Defence Economics*, vol. 2, no. 2 (1991), pp. 165–76.
[10] Wulf, H., 'Developing countries', eds N. Ball and M. Leitenberg, *The Structure of the Defence Industry: An International Survey* (St Martins Press: New York, 1983); Matthews, R., 'The development of the South African military industrial complex', *Defence Analysis*, vol. 4, no. 1 (1988), pp. 7–24; and Landgren (note 4).

**Table 2.4.** Arms production in the national economy, 1961–89
Figures are average annual rates of growth expressed as percentages.

|  | 1961–68 | 1968–77 | 1977–89 |
|---|---|---|---|
| GDP | 6.2 | 3.5 | 2.2 |
| Manufacturing output | 9.4 | 5.1 | 2.0 |
| Manufacturing employment | 5.7 | 3.1 | 1.0 |
| Arms production | 44.8 | 17.7 | 11.2 |
| Arms industry employment | 72.1 | 15.2 | 12.5 |
| Government expenditure | 10.4 | 6.3 | 2.6 |
| Military expenditure | 20.9 | 10.7 | 2.5 |

*Sources:* South African Reserve Bank, *Quarterly Bulletin* (South African Reserve Bank: Pretoria, various issues); and private communications with Armscor officials (arms production and arms industry employment).

Average annual growth in the value of domestic arms production was significantly higher than GDP growth and the growth in manufacturing output for all three periods between 1961 and 1989 (see table 2.4). More importantly, the arms industry continued to expand after 1977 when growth in GDP and manufacturing output was slowing. Growth in employment in the arms industry was significantly greater than the growth in total manufacturing employment for all three periods, and military expenditure grew faster than total government expenditure during 1961–68 and 1968–77.

*1961–68*

Before 1961 South Africa relied quite heavily on imports of complete weapon systems from abroad—among others from Britain and the USA.[11] Withdrawal from the Commonwealth in 1961 and the imposition of the UN arms embargo in 1963 severely limited South Africa's access to its traditional foreign sources of armaments. It also provided the impetus for a shift away from importing complete weapon systems towards the assembly and licensed production of various types of armaments.[12]

The shift away from importing arms towards licensed production and/or domestic procurement demanded substantial investment in training, technology, R&D and production equipment. It also necessitated a number of changes in the institutional structures for domestic arms production. The Armaments Production Board which was established in 1964 had a mandate for 'the manufacture, procurement and supply of all armaments and equipment for the SADF'.[13] The Board also took over the Department of Defence's workshops at Lyttleton and the ammunition section of the South Africa Mint, and was authorized to

[11] Landgren (note 4), pp. 39–40.
[12] Landgren (note 4), p. 41.
[13] Matthews (note 10), p. 8.

encourage and coordinate arms production in private companies. By the mid-1960s nearly 1000 private-sector firms were involved in arms production.[14]

In 1966 a Defence Council was established to advise the minister of defence on military-related R&D and armaments production issues. The nine members of the council included representatives from the SADF, the Council for Scientific and Industrial Research (CSIR), the Industrial Development Corporation (IDC), the chairman of the State Tender Board, the General Manager of the Armaments Production Board and a number of private-sector business leaders.[15] The Defence Council and the Armaments Production Board were the first institutionalized links between the state, the military and the private sector, and provided the institutional foundations for the domestic arms industry.[16]

By the late 1960s South Africa had reached stage 5 of the military industrialization process—final assembly of certain weapon systems (small arms, armoured cars and jet trainers) and some local component production (such as ammunition). Military industrialization between 1961 and 1968 was funded by sustained real increases in military expenditure and took place in the context of an economic boom, with average annual growth rates of over 6 per cent.[17] It also coincided with the expansion of South Africa's manufacturing sector, which was funded by large mineral and agricultural export earnings and supported by import substitution policies.

## 1968–77

In 1968 Armscor, the state-owned Armaments Development and Production Corporation, was set up. Between 1968 and 1977 South Africa's arms industry reached a critical stage in its development. During this period it began to undertake minor R&D improvements to local licence-produced armaments and started limited independent production of less sophisticated weapons. These developments were accompanied by important changes to the institutional structures which coordinated domestic arms production. As a result of South Africa's deteriorating internal and external security environment and the prospect of a strengthening of the UN arms embargo, the state became increasingly involved in the production of armaments, and systematic efforts were made to strengthen and expand the domestic arms industry in conjunction with the private sector.

The importation of finished weapon systems was increasingly substituted by the negotiation of licences to assemble and manufacture foreign weapons and by the importation of components and technologies needed for weapons produced in South Africa. During this period South Africa's major foreign

---

[14] South African Department of Defence, *White Paper on Defence and Armaments Production 1965–1967* (Government Printer: Pretoria, 1967).
[15] *White Paper on Defence and Armaments Production 1965–67* (note 14), p. 7.
[16] Philip (note 6), pp. 203–207.
[17] Moll, T., 'Did the apartheid economy fail?', *Journal of Southern African Studies*, vol. 17, no. 2 (1991), pp. 271–91.

sources of weapons and technology were France, Israel and Italy and to a lesser extent Belgium, Britain, the Federal Republic of Germany and the USA.[18]

In 1967 P. W. Botha, then Minister of Defence, visited armaments factories in Portugal and France as part of an in-depth investigation into various 'models' for arms production in South Africa.[19] In 1968 the Armaments Production Board's name was changed to the Armaments Board. It was tasked with the control, development, procurement and supply of armaments to the SADF, as well as ensuring the optimal utilization of the private sector for arms production. In the same year the IDC helped to establish Armscor. The objects and tasks of Armscor, as defined by the Armaments Development and Production Act no. 57 of 1968, were to 'promote and coordinate the development, manufacture, standardization, maintenance, acquisition, or supply of armaments . . . utilizing the services of any person, body or institution or any department of the state'.[20] As part of its mandate Armscor was tasked with controlling arms production, either through subsidiary companies or through companies in which it would participate by way of shareholdings and/or financial assistance.

Over the next few years the R&D and production facilities of the public-sector arms industry were rationalized and expanded. Armscor expanded its production activities by acquiring private companies and establishing new subsidiary companies. The Defence Ordnance Workshop and the Ammunition Section of the South Africa Mint, which were renamed the Lyttleton Engineering Works and Pretoria Metal Pressings, respectively, became the first full subsidiaries of Armscor.[21] In 1968 a missile test range was set up in northern Natal and a new subsidiary, Kentron, was established to work on the development of missile technology, previously carried out by the National Institute for Rocket Research (NIRR).[22]

In 1969 Armscor took over Atlas Aircraft Corporation, established in 1964 with government assistance, and Musgrave, a private firm manufacturing rifles and high-precision arms components. In the following year it took over two private-sector factories which manufactured a variety of propellants and explosives, later called Somchem and Naschem, respectively. Armscor also became the major shareholder in the private firm Ronden, which manufactured pyrotechnic products and later became known as Swartklip.[23] A new production facility, Eloptro, was set up in 1974 to manufacture optical and electro-optical equipment for various weapon systems.[24] In 1975 the Institute for Maritime Technology (IMT) was established in Simonstown to provide R&D support for the navy.

[18] Brzoska, M., 'Arming South Africa in the shadow of the UN arms embargo', *Defence Analysis*, vol. 7, no. 1 (1991), pp. 21–38.

[19] Frankel, P., *Pretoria's Praetorians* (Cambridge University Press: Cambridge, 1984), p. 82.

[20] Armaments Development and Production Act, Act no. 57, *Statutes of the Republic of South Africa* (Government Printer: Pretoria, 1968), pp. 420–21.

[21] 'Armscor', *SALVO*, special edition, 1987, p. 4.

[22] Cawthra, G., *Brutal Force: The Apartheid War Machine* (International Defence and Aid Fund for Southern Africa: London, 1986), p. 98.

[23] 'Armscor' (note 21), p. 4.

[24] 'Armscor' (note 21), p. 5.

During this period the government continued to import finished weapon systems and negotiated with overseas arms suppliers to assemble and produce weapons locally under licence that the domestic industry could not supply. In 1971 an agreement was reached with the French companies Dassault and Snecma which provided for the assembly and manufacture of the Mirage F-1 fighter aircraft in South Africa under licence at Atlas Aircraft Corporation.[25]

By the early 1970s the domestic arms industry had reached stage 6 of the military industrialization process and was able to undertake complete licensed production of certain weapon systems. For example, between 1968 and 1977 Atlas Aircraft assembled under licence from Aermacchi of Italy the Impala I and II jet trainers. These aircraft were powered by the British Rolls Royce Viper engine, which was licensed by Britain to the Italian firm Piaggio, which in turn sub-licensed it to South Africa.[26]

During this period South Africa also attempted to initiate a strategy of common arms production within the Southern African region with Rhodesia, a number of independent African states and Portugal. This regional military industrialization plan proceeded without much success until the collapse of the Salazar regime in Portugal in 1974. There was a high level of cooperation with the Rhodesian arms industry after 1968, which allowed South African engineers the chance to get direct experience of wartime arms production and development.[27] This industrial cooperation was concentrated in the armoured vehicle industry and resulted in the development of an entire series of armoured personnel carriers (including the Hippo) and military transport vehicles (such as the Samil) which were produced in the mid-1970s.[28]

Increasing international opposition to apartheid and worldwide demands for a mandatory arms embargo against South Africa prompted the government to embark on a major reorganization and expansion of the arms industry during the mid-1970s. This expansion necessitated significant increases in the defence budget: it grew at an average of 10 per cent per annum between 1968 and 1977. The government decided that the procurement of armaments and control of domestic arms production would be best served by one body, and during 1975 and 1976 the Armaments Board and the Armaments Development and Production Corporation were amalgamated to form the 'new' Armscor (now the Armaments Corporation of South Africa). With an initial share capital of 100 million rand, it assumed responsibility for the procurement, development and manufacture of armaments as well as overall coordination of the local arms industry.[29]

In 1977 with UN Security Council Resolution 418 came the mandatory arms embargo against South Africa. Cobbett suggests that its imposition was largely

[25] Landgren (note 4), pp. 70–73.
[26] Väyrynen, R., 'The role of transnational corporations in the military sector of South Africa', *Journal of Southern African Affairs*, vol. 5, no. 2 (1980), pp. 199–255.
[27] Landgren (note 4), p. 82.
[28] *Defence Today*, no. 89–90 (1985), p. 412.
[29] The restructuring and expansion of Armscor was also funded by a secret government grant of 1200 million rand. Landgren (note 4), p. 42.

a result of South Africa's (abortive) invasion of Angola in 1975, the government's brutal suppression of the Soweto uprisings in 1976 and the death of Steve Biko in prison in 1977.[30] When the embargo was imposed, however, the South African arms industry was already either producing or in the process of acquiring the knowledge to produce a wide spectrum of armaments.[31] It had reached stage 7—limited R&D improvements to local licence-produced products—and certain sections of the industry (ammunition and small arms) were close to reaching stage 8—limited independent production of some less sophisticated armaments as well as limited production of more advanced armaments such as aircraft, ships and missiles. The SADF's involvement in the Rhodesian war from the mid-1960s until the late 1970s and the fact that the Rhodesian security forces used mainly South African-produced equipment also contributed to the increasing independence of the South African arms industry.[32]

These developments in South Africa's military industrialization took place at a time of sustained economic growth between 1968 and 1973 and then of slower growth between 1974 and 1977,[33] and in a context of increasing import substitution.[34] By the late 1970s import substitution was more or less complete in consumer goods and significant import substitution had occurred in sectors such as basic chemicals, metal products, motor vehicle parts and non-ferrous metals, although in some sectors, such as transport equipment and machinery, significant import penetration or negative import substitution had taken place and import-penetration ratios were still high.[35]

### 1977–89

Between 1977 and 1989 the South African arms industry expanded considerably in response to the imposition of the UN arms embargo and South Africa's increasing involvement in a number of regional conflicts. To this end, R&D expenditure was dramatically increased (see table 2.13), new state-owned R&D and production facilities were established, and the private sector became increasingly involved in defence production as contractors, subcontractors and suppliers of military equipment and technology.

Between 1977 and 1981 Armscor's R&D and production activities were expanded and reorganized. Its assets and employment more than doubled between 1977 and 1982. In the late 1980s new R&D and testing facilities such as the Advena Central Laboratories and the Overberg Testing Range were established and two new subsidiary companies, Infoplan (computers) and Houwteq (missile systems), were established. By the end of the 1980s the

---

[30] Cobbett, W., 'Apartheid's army and the arms embargo', eds Cock and Nathan (note 1), p. 235.

[31] Brzoska (note 9), p. 205.

[32] Landgren (note 4), p. 177.

[33] Moll (note 17).

[34] Holden, M., 'The choice of trade strategy', eds N. Nattrass and E. Ardington, *The Political Economy of South Africa* (Oxford University Press: Cape Town, 1990), pp. 260–74; and Kahn, S., 'Import penetration and import demand in the South African economy', *South African Journal of Economics*, vol. 55, no. 3 (1987).

[35] Holden (note 34), p. 263.

Armscor organization included 12 subsidiary companies and a number of R&D and testing facilities.

In addition to investing in arms production capabilities the government pursued self-sufficiency in a number of related industries, such as engines, gear-boxes and axles. Using German (Daimler Benz) and British (Perkins Diesel) engine technology and equipment, the Atlantis Diesel Engines (ADE) plant was set up near Cape Town in the early 1980s. The plant cost 350 million rand and was financed by the IDC, and in the following years almost all SADF armoured vehicles and transport vehicles were fitted with ADE engines.[36] To cover the costs of establishing the ADE plant, the government raised import tariffs on diesel engines to a prohibitive level, despite strong protests from the private sector.[37] During the mid-1980s the government also invested heavily in the electronics sector, and South Africa's first microchip plant was established with West German assistance in 1984.[38]

During the early 1980s South Africa's arms industry began to experience a number of economic problems as a result of increasing production costs (the rising costs of embargoed inputs), excess capacities and declining domestic demand. Drastic staff cuts were made at Armscor and employment declined from 26 000 in 1981 to 23 000 in 1985. A number of defence contracts with private-sector firms were cancelled, some of Armscor's production activities were rationalized and a government commission was set up in 1984 to investigate the corporation's financial problems.[39] These problems, against the background of deteriorating economic performance, were exacerbated by the fact that the UN embargo forced Armscor to adopt a number of uneconomic practices—to invest in R&D and production facilities which were commercially unviable, tool up for short production runs, and stockpile up to four years' supply of certain items which were not available in South Africa.[40]

Arms exports were seen as a solution to the problem of surplus capacity, while at the same time helping to preserve accumulated skills and technologies.[41] Legislation covering the export of arms was passed and a new international sales and marketing organization, Nimrod, was created in 1982 within Armscor to handle arms exports.[42] South Africa's appearance at an international defence exhibition in Greece in 1982 marked its 'official' entry into the international arms market, and over the following years it embarked on an aggressive international sales drive targeted specifically at developing country

---

[36] Landgren (note 4).

[37] Grundy, K., *The Rise of the South African Security Establishment*, Bradlow Paper no. 1 (South African Institute for International Affairs: Johannesburg, 1983), p. 21.

[38] Brzoska (note 18), p. 26.

[39] South African Department of Defence, *White Paper on Defence and Armaments Supply 1984* (Government Printer: Pretoria, 1984), p. 19.

[40] Landgren (note 4), p. 58.

[41] Geldenhuys, D., *Isolated States: A Comparative Analysis* (Cambridge University Press: Cambridge, 1990). He suggests that South Africa's decision to export arms may also have been politically motivated in the attempt to use 'arms diplomacy' in the same way as Israel to gain access to states that were formally 'closed' to South Africa.

[42] Landgren (note 4), p. 182.

markets. Its entry into the international arms market coincided with changes in its trade regime and benefited from government attempts to promote manufactured exports in general. Following the recommendations of the Reynders Commission (1972) and the Van Huyssteen Committee (1978), the government introduced a number of incentives in the 1980s to promote exports, such as direct cash subsidies, tax concessions, rail freight concessions and rebates on import duties on inputs.[43]

During the 1980s the local arms industry invested large amounts of resources in an effort to reach the upper stages of the military industrialization process— the ability to design, develop and manufacture a wide range of weapon systems independently. This push for self-sufficiency and the need to develop 'new' and improved weapon systems was contingent upon a number of factors such as the strengthening of the UN arms embargo, attempts to break into the international arms market, and the ongoing involvement in the conflict in Angola and Namibia. Against this background a number of 'new' indigenous weapon systems were presented by the local arms industry during the 1980s. Most of these were either upgrades or modifications of existing weapons and still dependent to some extent on imported components, foreign technology or foreign designs.[44]

By the end of the 1980s the arms industry as a whole had reached stage 9— independent R&D and production of less sophisticated weapons and limited R&D and production of more advanced weapons. Certain sectors, such as ammunition, artillery, small arms and armoured vehicles, had reached stage 10 while others, such as military aircraft, missiles and naval vessels, were at much lower stages.[45] Landgren suggests that by the end of the 1980s the South African arms industry had not reached the stage of independent R&D and production of weapons but had achieved a capacity to retrofit, redesign or upgrade weapon systems.[46] Brzoska suggests that, because of limited R&D resources and the UN arms embargo, the local arms industry did not waste time and money trying to reproduce or emulate the R&D which had already been carried out in the major Western arms producers.[47] Instead it concentrated on acquiring a capacity for upgrading, add-on and add-up engineering. These three modes of production were extensively used in the 1980s to modernize and upgrade existing weapon systems, to develop 'new' weapon systems and to meet the changing equipment needs of the SADF.[48]

The considerable expansion of the domestic arms industry between 1977 and 1989 occurred at a time of economic slow-down and increasing international

---

[43] Holden, M., 'Trade policy, income distribution and growth', ed. R. Schrire, *Wealth or Poverty? Critical Choices for South Africa* (Oxford University Press: Cape Town, 1992), p. 319.

[44] Brzoska (note 18), p. 25.

[45] Information based on personal communications with Armscor officials; C. Foss, Military Technology Editor, *Jane's Defence Weekly*; and H. Heitman, South Africa correspondent, *Jane's Defence Weekly*. See also Anthony, I., 'The third-tier countries: production of major weapons', ed. H. Wulf, SIPRI, *Arms Industry Limited* (Oxford University Press: Oxford, 1993).

[46] Landgren (note 4), p. 14.

[47] Brzoska (note 18), p. 25.

[48] Landgren (note 4); and Brzoska (note 18), p. 25.

economic isolation. Although the economy experienced positive growth between 1979 and 1981 as a result of high gold prices and the primary commodity boom which accompanied the oil price crisis, the period 1977–89 was characterized by slower economic growth.

## III. Self-sufficiency

By the late 1980s South Africa's arms industry had achieved a relatively high level of self-sufficiency in terms of being able to supply the SADF with most of its equipment requirements. Although some of the weapon systems were still produced under (terminated) licensing agreements, and although certain components and materials had to be imported, the local industry had in effect replaced imported weapon systems with domestically-produced products. The degree of self-sufficiency varied as between sectors of the arms industry: some, such as ammunition and armoured vehicles, were practically self-sufficient, whereas others, such as military aircraft, still relied to a large extent on foreign inputs of technology or components.

After 1977 Armscor and the government made a number of public statements concerning the degree of South Africa's self-sufficiency in arms production. In 1977 P. W. Botha told parliament that 57 per cent of South Africa's arms and military equipment came from overseas sources. In 1982 the Chairman of Armscor stated that 15 per cent of South Africa's defence budget was spent abroad.[49] Grundy suggested that in the mid-1980s 15 per cent of the defence budget was spent on material imports.[50] A study in 1988 reported that Armscor only imported 5 per cent of its requirements, as opposed to 70 per cent before the arms embargo began.[51] According to information recently obtained from Armscor, South Africa's arms import ratio (direct and indirect imports as a percentage of total procurement expenditure) fell from nearly 100 per cent in the early 1960s to 42 per cent in the late 1980s. The import ratio of 42 per cent in the late 1980s included roughly 30 per cent on direct imports, the other 12 per cent being indirect imports (foreign content of locally supplied equipment).[52]

One of the conditions of South Africa's becoming increasingly self-sufficient in arms production was its ability to continue to obtain foreign inputs by circumventing the UN embargo. A number of studies have highlighted the ways in which it did this.[53] The arms industry was 'forced' to adopt a number of covert and illegal practices (such as smuggling and front companies) in order to obtain much-needed inputs, including technology and components. The many multinational companies in South Africa, particularly in the electronics and motor vehicle sectors, could act as 'conduits' for the acquisition of foreign tech-

[49] Leonard, R., *South Africa at War* (Laurence Hill: Westport, Conn., 1983), p. 140.
[50] Grundy, K., *The Militarisation of South African Politics* (I. B. Tauris: London, 1986), p. 46.
[51] *South Africa Country Report, no. 2* (Economist Intelligence Unit: London, 1988), p. 14.
[52] Private communication with Armscor official, Pretoria, Apr. 1994.
[53] Väyrynen (note 26); Landgren (note 4); and Brzoska (note 18).

nology and components.[54] According to Landgren the dual-use nature of much of this technology, particularly electronics technology, meant that it could be acquired without contravening the UN embargo.[55] The government also played an important role in assisting local companies to acquire the local subsidiaries of foreign companies, particularly in those sectors such as electronics which were strategically important for the arms industry.[56] In addition the government introduced various legislative measures, such as the National Supplies and Procurement Act of 1970 (amended in 1979) and the Protection of Business Act of 1978, which allowed the minister of defence to order any firm in South Africa, including foreign-owned firms, to sell or produce goods for the government (i.e., the SADF).[57]

## IV. The size and structure of the domestic arms industry

The arms industry occupied a privileged position in terms of access to state resources because of the political and strategic factors which prompted its development and expansion. Furthermore, the UN arms embargo meant that the South African state was forced to use the private sector's capabilities and facilities in order to source critical inputs and to meet the ever-expanding armament needs of the SADF. By the end of the 1980s the arms industry represented one of the most significant of the country's manufacturing sectors and was spread between the public and private sectors.

### The public-sector arms industry

The public-sector arms industry in South Africa was entirely concentrated in one company, Armscor, which after 1977 functioned as an armaments procurement and production organization. Armscor, together with the SADF, was part of the Department of Defence and was directly accountable to the Minister of Defence. Because of the nature of its activities, Armscor was also a member of the Defence Planning Committee, the highest-level defence planning structure in the country. Its Board of Directors, appointed by the president, was directly responsible to the Minister of Defence for all matters pertaining to the activities of Armscor while the Management Board was charged with the day-to-day running of the corporation.[58] Armscor's production and research activities were located in a number of subsidiary companies and research and testing facilities in various parts of the country (see table 2.5).

[54] Väyrynen (note 26). In 1980 it was estimated that there were between 2000 and 2500 multinational companies in South Africa, although this number had declined to around 1000 by the mid-1980s. Landgren (note 4), pp. 22–23.

[55] Landgren (note 4), pp. 233–36.

[56] Väyrynen (note 26); and Landgren (note 4).

[57] Simpson (note 7), p. 225.

[58] South African Department of Defence, Briefing on the Organisation and Functions of the South African Defence Force and the Armaments Corporation of South Africa Ltd 1987 (Government Printer: Pretoria, 1987), pp. 54–59.

**Table 2.5.** The structure of Armscor, 1989

| Subsidiary/facility | Products/services | Location |
|---|---|---|
| Head Office | Administration | Transvaal |
| *Subsidiaries* | | |
| Atlas Aircraft | Military aircraft | Transvaal |
| Telcast | High-technology alloys | Transvaal |
| Eloptro | Electro-optics | Transvaal |
| Infoplan | Computer services | Transvaal |
| Kentron | Missiles/guided weapons | Transvaal |
| Lyttleton Engineering | Small arms/artillery | Transvaal |
| Musgrave | Rifles/shotguns | Orange Free State |
| Naschem | Large-calibre ammunition | Transvaal |
| Pretoria Metal Pressings | Small-calibre ammunition | Transvaal |
| Somchem | Propellants/explosives | Western Cape |
| Swartklip | Pyrotechnics/grenades | Western Cape |
| Houwteq | Missile systems | Western Cape |
| *Facilities* | | |
| Advena | Research laboratory | Transvaal |
| Inst. for Maritime Technology | Naval research | Western Cape |
| Overberg Test Range | Missile test range | Western Cape |
| Alkantpan | Ballistic test range | Northern Cape |
| Paardefontein | Antenna test facility | Transvaal |
| St Lucia | Missile and artillery test range | Natal |
| Milistan | Systems analysis | Transvaal |
| Gennan Systems | Engineering | Transvaal |
| Gerotek | Vehicle test range | Transvaal |

*Source:* Private communications with Armscor officials.

As a result of its access to state resources, by 1989 Armscor had developed into the 30th largest company in the country in terms of total assets, 15th in terms of total employment,[59] the 34th largest arms company in the world and the largest in a developing country.[60] Its total assets increased from 546 million rand to over 2 billion rand between 1968 and 1989, while employment rose by a factor of 10 during the same period (see table 2.6). Its capital : labour ratio, as measured by total assets per employee, was significantly higher than that of private-sector defence companies such as Altech, Dorbyl and Reunert.[61]

[59] *Top Companies* (Financial Mail: Johannesburg, 22 June 1990), p. 124.
[60] Miggiano, P., Sköns, E. and Wulf, H., 'Arms production', *SIPRI Yearbook 1992: World Armaments and Disarmament* (Oxford University Press: Oxford, 1992), pp. 361–97.
[61] Armscor's capital : labour ratio in 1989 was 142 700 rand. The average for the manufacturing sector was 63 000 rand. Figures derived from official Armscor figures and the *South African Statistics Yearbook 1994* (Central Statistics Service: Pretoria, 1994), pp. 12.40–12.45.

**Table 2.6.** Armscor assets and employment, 1968–89

|  | 1968 | 1970 | 1975 | 1977 | 1980 | 1985 | 1989 |
|---|---|---|---|---|---|---|---|
| Total assets (m. rand, constant 1985 prices) | 546 | 743 | 1 090 | 1 185 | 1 951 | 1 590 | 2 162 |
| Employment | 2 919 | 3 143 | 7 390 | 10 590 | 24 560 | 23 310 | 26 348 |

*Source:* Private communications with Armscor officials.

## The private-sector arms industry

The size and structure of the private-sector arms industry during the apartheid era are more difficult to describe because South African economic data (in common with those of all other countries) do not have a separate category for armaments in the standard industrial classification. Thus there is no official information on the number of private-sector firms involved in arms production, the relative distribution of arms-production activities in the various sectors and sub-sectors of the manufacturing sector or the value and volume of arms-production activities. During the mid-1980s Armscor estimated that about 70 per cent of domestic defence production was carried out by private-sector firms and 30 per cent by the Armscor subsidiary companies.[62]

The number of private-sector firms involved in domestic arms production between 1961 and 1989 fluctuated according to demand and according to the equipment requirements of the SADF.[63] Armscor stated in the mid-1980s that it had contracts with 2271 private-sector firms, of which 1083 (48 per cent) were direct contractors and 1188 (52 per cent) were suppliers of standard items (see table 2.8). In 1984 it was estimated that approximately 400 private-sector firms were totally dependent upon Armscor contracts for their survival.[64] The large number of firms which acted as contractors and subcontractors to Armscor (over 1000) was a function of the low level of vertical integration in the private-sector arms industry. The private-sector firms which were contractors or suppliers to Armscor in the late 1980s represented over 10 per cent of firms in the manufacturing sector.[65]

The structure of the private-sector arms industry in the late 1980s reflected the structure and ownership patterns of South Africa's manufacturing sector. Large and highly diversified corporate groups were a common feature of the manufacturing sector and were in turn either owned or controlled (majority stake) by one or more of the six large financial–mining–industrial conglom-

[62] 'Armscor: production strategy and industrial suppliers', *SALVO*, vol. 13, no. 10 (1984), p. 27.
[63] South African Department of Defence, *White Paper on Defence and Armaments Supply 1979* (Government Printer: Pretoria, 1979).
[64] Ratcliffe, S., 'Forced relations: the state, the crisis and the rise of militarism in South Africa'. Unpublished Honours thesis, University of the Witwatersrand, Johannesburg, 1983, p. 77.
[65] 1988 Manufacturing Census, *South African Statistics Yearbook 1994* (Central Statistics Services: Pretoria, 1994), pp. 12.40–12.45.

**Table 2.7.** The top industrial companies, 1989
Companies are ranked by total assets.

| Company | Industrial sector | Total assets (current m. rand) | Employment | Capital : labour ratio[a] |
|---|---|---|---|---|
| *Public sector* | | | | |
| Iscor | Steel | 7 791 | 58 000 | 134.3 |
| Sasol | Chemicals | 6 352 | 29 300 | 216.8 |
| Armscor | Armaments | 3 761 | 26 348 | 142.7 |
| *Private sector* | | | | |
| Sappi | Paper | 4 623 | 22 033 | 209.8 |
| AECI | Chemicals | 3 143 | 27 300 | 115.1 |
| Tongaat | Food | 2 346 | 40 201 | 58.4 |
| Nampak | Paper | 1 855 | 22 627 | 81.9 |
| Sentrachem | Chemicals | 1 719 | 6 625 | 259.5 |
| Hiveld | Steel | 1 534 | 7 593 | 202.1 |
| Dorbyl | Engineering | 1 383 | 23 371 | 59.2 |
| Afrox | Engineering | 877 | 6 971 | 125.8 |
| Reunert | Electronics | 784 | 13 309 | 58.9 |
| ICS | Food | 765 | 13 794 | 55.5 |
| Haggie | Engineering | 726 | 9 051 | 80.2 |
| Altech | Electronics | 551 | 5 500 | 100.2 |
| Grintek | Electronics | 509 | 3 453 | 147.4 |

[a] Total assets per employee in thousand current rand.

*Source: Top Companies* (Financial Mail: Johannesburg, 22 June 1990).

erates—Anglo-American, Anglovaal, Liberty Life, Old Mutual, the Rembrandt Group and Sanlam. These six controlled approximately 90 per cent of the asset value of the Johannesburg Stock Exchange in the later 1980s.[66] The private-sector arms industry was dominated by three large industrial groups publicly quoted on the Johannesburg Stock Exchange—Altech, Grintek and Reunert. All three were in turn either owned or controlled by one of the six large conglomerates. Altech was controlled by Anglo-American, Grintek by Anglovaal and Reunert by Old Mutual. Each of the three groups was made up of a number of divisions and subsidiary companies, and collectively they dominated most of the sectors and sub-sectors of the local defence market.

In terms of the geographical distribution of private-sector defence firms, more than 75 per cent were located in the Pretoria–Witwatersrand–Vereeniging (PWV) region.[67] This corresponded to the geographical distribution of the public-sector arms industry: Armscor and most of its subsidiaries and facilities were located there. The PWV also had the dominant regional economy and the highest concentration of high-technology firms in the manufacturing sector. The

[66] For a discussion of the ownership structure of the South African manufacturing sector, see Joffe, A., Kaplinsky, R., Kaplan, D. and Lewis, D., *Improving Manufacturing Performance in South Africa* (University of Cape Town Press: Cape Town, 1995), pp. 133–86.
[67] *The Armaments Industry in South Africa* (Armscor: Pretoria, 1984), p. 7.

**Table 2.8.** Private-sector defence contractors, 1984

| Sector | No. of firms | Share of total (%) |
|---|---|---|
| Electrical engineering | 211 | 9.2 |
| Chemical engineering | 120 | 5.2 |
| Heavy engineering | 106 | 4.6 |
| Light industry | 621 | 27.3 |
| Shipbuilding and others | 25 | 1.1 |
| Sub-total | 1 083 | 47.7 |
| Delivery off the shelf[a] | 1 188 | 52.3 |
| **Total** | **2 271** | **100.0** |

[a] Purchase by Armscor of standard items of companies' output, e.g., nuts, bolts and springs.
*Source: The Armaments Industry in South Africa* (Armscor: Pretoria, 1984), p. 7.


Western Cape region and the Durban–Pietermaritzburg area were the only other regional economies with a significant number of private-sector defence firms.[68]

## V. The economic significance of the domestic arms industry

The literature on arms production in developing countries highlights the fact that establishing a domestic arms industry can have both economic costs and benefits.[69] The benefits relate to employment creation, human capital development, foreign-exchange savings and technological spin-offs. The costs relate to the dampening of employment creation because of the capital intensity of arms production and the absorption of scarce resources, both of which contribute to supply-side shortages and the crowding out of civilian investment.

### Employment

The development and expansion of South Africa's arms industry had a significant impact on employment creation. Employment in the industry increased from less than 1000 in 1961 to over 130 000 in 1989. As a percentage of total manufacturing employment it rose from 0.1 per cent in 1961 to 8.3 per cent in 1989, and as a percentage of total employment in the non-agricultural sector from less than 0.1 per cent in 1961 to 2.3 per cent in 1989 (see table 2.9). Total arms industry employment and its share in total manufacturing employment

---

[68] For a study of the geographical distribution of South Africa's arms industry, see Rogerson, C., 'Defending apartheid: Armscor and the geography of military production in South Africa', *GeoJournal*, vol. 22, no. 3 (1990), pp. 241–50.

[69] For a discussion of the economic costs and benefits of arms production in developing countries, see chapter 1 in this volume. See also Dunne, P., 'The defence industrial base', eds K. Hartley and T. Sandler, *Handbook of Defence Economics, Vol. 1* (Elsevier: Amsterdam, 1995); and Wulf, H., 'Arms production in third world countries: effects on industrialisation', ed. C. Schmidt, *The Economics of Military Expenditure* (Macmillan: London, 1987).

**Table 2.9.** Arms industry employment, 1961–89

|  | 1961 | 1965 | 1970 | 1975 | 1980 | 1985 | 1989 |
|---|---|---|---|---|---|---|---|
| Total non-agric. employment (th.) | 2 643.00 | 2 933.90 | 3 741.30 | 4 576.80 | 4 893.20 | 5 246.30 | 5 650.20 |
| Total manufacturing employment (th.) | 673.00 | 913.50 | 1 070.00 | 1 312.70 | 1 460.00 | 1 484.00 | 1 583.30 |
| Total arms industry employment (th.) | 0.83 | 9.49 | 15.72 | 36.95 | 122.80 | 116.55 | 131.75 |
| Arms ind. share of total employment (%) | 0.1 | 0.3 | 0.4 | 0.8 | 2.5 | 2.2 | 2.3 |
| Arms ind. share of manufact. employ't (%) | 0.1 | 1.0 | 1.5 | 2.8 | 8.4 | 7.9 | 8.3 |

*Sources:* South African Reserve Bank, *Quarterly Bulletin* (South African Reserve Bank: Pretoria, various issues); and private communications with Armscor officials (total arms industry employment).

increased significantly from the mid-1980s as a result of South Africa's increasing military involvement in Angola and Namibia and the use of the SADF for internal military purposes. In many other manufacturing sectors growth in employment virtually came to a halt during the 1980s.[70]

While the arms industry emerged as a significant provider of jobs during the 1970s and 1980s, most of the jobs created were highly capital- and skill-intensive and in most cases reserved for whites. The arms industry thus not only absorbed scarce resources of skilled labour, but perpetuated the dualist (racist) structure of the labour market.[71] The employment creation benefits of arms production must be viewed in the light of the opportunity costs associated with creating jobs which were inappropriate for South Africa's factor endowments (scarce capital and an abundance of unskilled labour), and which made little or no significant contribution to the performance of the manufacturing sector.

### Investment

Many empirical studies have highlighted the fact that increases in military expenditure can have a negative impact on levels of non-military spending, and that investment in domestic arms production may have negative resource allocation effects by crowding out civilian investment, which can have adverse effects on a country's economic growth and international competitiveness.[72] The large shares of net investment in strategic industries (such as armaments) in

---

[70] On the trends in manufacturing employment during the 1980s, see Joffe *et al.* (note 66), p. 7.

[71] The capital intensity of the arms industry in the late 1980s was significantly higher than that of most other industrial sectors, with the exception of chemicals and iron and steel, and more than double the average for the manufacturing sector as a whole. See note 61.

[72] Hartley, K. and Sandler, T., *The Economics of Defence* (Cambridge University Press: Cambridge, 1995).

**Table 2.10.** Arms imports, 1970–89

|  | 1970 | 1975 | 1980 | 1985 | 1989 |
|---|---|---|---|---|---|
| Arms imports (m. rand, constant 1985 prices) | 709 | 1 576 | 847 | 800 | 1 407 |
| Total imports (m. rand, constant 1985 prices) | 23 799 | 31 023 | 31 839 | 28 409 | 31 077 |
| Arms as share of total imports (%) | 3.0 | 5.1 | 2.7 | 2.8 | 4.5 |
| Manufactured imports (m. rand, constant 1985 prices) | 19 985 | 23 153 | 20 785 | 19 645 | 23 600 |
| Arms as share of manufactured imports (%) | 3.5 | 6.8 | 4.1 | 4.1 | 6.0 |
| Import ratio[a] | 81.0 | 66.0 | 47.8 | 42.0 | 42.0 |

[a] Direct and indirect military imports as a percentage of total procurement expenditure.

*Sources:* South African Reserve Bank, *Quarterly Bulletin* (South African Reserve Bank: Pretoria, various issues); Central Economic Advisory Service (1994); and private communications with Armscor officials (import ratio and arms imports).

South Africa in the 1970s and 1980s often led to under-investment in the country's most labour-intensive sectors (e.g., clothing, leather/footwear and wood/furniture). Kaplinsky suggests that if there had been complementary investments in the more labour-intensive 'downstream' ends (such as plastics) of sectors such as chemicals, or in the very labour-intensive sectors such as clothing, then the growth of manufacturing output during the 1980s would have been greater.[73] Investment in arms production is also generally assumed to be less productive than civilian investment because the primary aim of military production is product innovation—to raise the performance of the weapon system—rather than process innovation—to reduce costs through new technology and higher factor productivity.[74]

Some studies have highlighted the fact that the excessive share of total investment in strategic industries (such as armaments) in South Africa in the 1970s and 1980s was a form of 'misinvestment': vast resources were invested in the 'wrong' sectors and/or sub-sectors.[75] This misinvestment also involved wrong choices of technology (such as synthetic fuels), which were driven by strategic considerations (such as embargoes) rather than by relative factor prices.

### Foreign-exchange savings

One of the major reasons why developing countries establish domestic arms industries is to save on foreign exchange. However, some studies have suggested that, while establishing a domestic arms industry may reduce the value of direct arms imports, the value of indirect imports (including technology,

[73] Kaplinsky, R., 'South African industrial performance and structure in a comparative context', Paper presented at the Industrial Strategy Meeting, Johannesburg, 6–10 July 1992.

[74] Brzoska, M., 'The impact of arms production in the third world', *Armed Forces and Society*, vol. 15, no. 4 (1989), pp. 507–30.

[75] Kaplinsky (note 73); and Joffe *et al.* (note 66).

capital goods and personnel) needed for domestic arms production may in fact exceed the foreign-exchange savings achieved by the reduction in direct arms imports.[76]

By the end of the 1980s, it was estimated that South Africa was still spending 1.4 billion rand (in 1985 prices) on direct and indirect (foreign content of locally supplied goods) military imports. While this was significantly less as a share of total procurement expenditure than in the 1970s before the imposition of the arms embargo, it still constituted a significant share of manufactured imports and total imports (see table 2.10).

The value of South Africa's arms imports and their share in total imports and manufactured imports peaked in the years before the imposition of the UN arms embargo in 1977 and then declined quite substantially before increasing in the late 1980s. The arms import ratio also declined significantly after the imposition of the embargo. However, these figures do not include the technology, capital goods and personnel which were needed for domestic arms production. If these are included, then the foreign-exchange costs of domestic arms production are much higher than the value of direct and indirect arms imports.

Earnings from arms exports can be used to offset the foreign-exchange costs of domestic arms production. The value of South Africa's arms exports rose from 33 million rand in 1982 to 347 million rand in 1987, dropping to 118 million rand in 1989, and averaging about 163 million rand in constant 1985 prices between 1982 and 1989—significantly less than the estimated average value of $300 million (at 1985 prices) of Brazil's arms exports during the same period.[77] Arms exports amounted to 3.1 per cent of manufactured exports and 0.9 per cent of total exports by value in 1987, dropping in 1989 (see table 2.11).

The relative 'success' of South Africa's arms export drive during the 1980s was related to a number of factors. They were relatively cheap (due to the low value of the rand) and robust, they had been developed with the needs of a fairly small, highly mobile defence force in mind, and they had been tested in and exposed to operational conditions (for instance, in South Africa's operations in Angola and Namibia).[78] Some commentators have argued that this success was also related to the fact that the country emerged as a 'willing exporter to a number of nations which for one reason or another [were] cut off from military business with the leading Western arms producers'.[79]

The export performance of the arms industry as measured by the share of exports in total production fluctuated between nearly 2 per cent and over 20 per cent between 1982 and 1989, with an average of just over 10 per cent (see table 2.11). This was significantly lower than the average export performance for the manufacturing sector as a whole, but higher than that of certain sectors

[76] Ball, N., *Security and Economy in the Third World* (Princeton University Press: Princeton, N.J., 1988); and Brauer (note 9).

[77] Franko-Jones, P., *The Brazilian Defence Industry* (Westview Press: Boulder, Colo., 1992), pp. 140–41.

[78] Heitman, H., 'South Africa's defence industry: challenges and prospects', *South African Defence Review*, no. 2 (1992), pp. 26–34.

[79] Landgren (note 4).

**Table 2.11.** Arms exports, 1982–89

|  | 1982 | 1983 | 1984 | 1985 | 1986 | 1987 | 1988 | 1989 |
|---|---|---|---|---|---|---|---|---|
| Arms exports (m. rand, constant 1985 prices) | 33 | 35 | 56 | 282 | 287 | 347 | 178 | 118 |
| Total exports (m. rand, constant 1985 prices) | 32 651 | 29 895 | 32 770 | 39 973 | 39 737 | 37 302 | 37 598 | 38 113 |
| Arms as share of total exports (%) | *0.1* | *0.1* | *0.2* | *0.7* | *0.7* | *0.9* | *0.5* | *0.3* |
| Manufact'd exports (m. rand, constant 1985 prices) | 9 330 | 7 788 | 8 845 | 12 122 | 12 903 | 11 239 | 12 177 | 13 205 |
| Arms as share of manufact'd exports (%) | *0.4* | *0.4* | *0.6* | *2.3* | *2.2* | *3.1* | *1.5* | *0.9* |
| Exports as share of total arms prod. (%) | *1.9* | *2.0* | *3.5* | *20.3* | *19.6* | *18.3* | *10.8* | *5.7* |

*Sources:* South African Reserve Bank, *Quarterly Bulletin* (South African Reserve Bank: Pretoria, various issues); Central Economic Advisory Service (1994); and private communications with Armscor officials (total arms exports).

such as motor vehicles, electrical machinery and metal products. The contribution of arms exports to the national economy, already fairly insignificant in terms of share of total exports, is even less significant if the costs of export subsidies and marketing (paid for by Armscor) are included, and if allowance is made for the fact that much of the R&D and production costs of export products was subsidized by the procurement budget.[80]

South Africa's trade balance in arms and military technology was negative between 1982 and 1989. Thus domestic arms production remained a net user of foreign exchange throughout the 1980s (see table 2.12).

## Technology transfer and spin-offs

One of the purported economic benefits of domestic arms production is technology transfer and spin-offs from the military to the civilian economy. Spin-offs may take the form of products, processes, organizational techniques and/or knowledge, and can enhance the technological and innovatory capabilities of the civilian manufacturing sector.

The existing evidence suggests that South Africa's domestic arms industry did not have a significant impact on the technological capabilities of the manu-

---

[80] Willett, S. and Batchelor, P., *To Trade or Not to Trade: The Costs and Benefits of South Africa's Arms Trade*, Working Paper no. 9 (Military Research Group: Johannesburg, 1994).

**Table 2.12.** The trade balance in armaments, 1982–89

Figures are in m. rand in constant 1985 prices.

|         | 1982 | 1983 | 1984 | 1985 | 1986 | 1987  | 1988  | 1989   |
|---------|------|------|------|------|------|-------|-------|--------|
| Imports | 721  | 781  | 810  | 800  | 797  | 1 089 | 1 076 | 1 407  |
| Exports | 33   | 35   | 56   | 282  | 287  | 347   | 178   | 118    |
| Balance | − 88 | − 746| − 754| − 518| − 510| − 742 | − 898 | − 1 289|

*Source:* Private communications with Armscor officials.

facturing sector through technology transfers or spin-offs.[81] The spin-offs that did occur were largely confined to the fields of electronics, explosives and the steel and metal industries.[82] Part of the reason for the lack of spin-off was the highly secretive nature of the arms industry, which prevented the sharing or transfer of technologies to the civilian sector.[83] In many cases the technology that was used for arms production was too expensive and/or too sophisticated for the production of commercial goods. Defence firms were in most cases sheltered from market forces and used technologies and production processes which were not commercially viable. In addition, Armscor was prohibited under the Armaments Development and Production Act of 1968 (as amended) from using its production facilities for commercial use, and this also prevented the realization of spin-offs to the civilian sector.

Expenditure on military R&D may stimulate civilian R&D and thereby enhance a country's technological performance and have a positive indirect effect on productivity growth. However, it may also depress or crowd out expenditure on non-military R&D and thus have a negative impact on a country's technological performance and international competitiveness.[84]

Military R&D began in earnest in South Africa in 1961. In the early 1970s a Defence Research Committee (DRC) was established to coordinate and manage military R&D programmes; in 1978 a new structure for coordinating military R&D, the Defence Research and Development Council (DRDC), supported by a technical secretariat, was established in the office of the Chief of Staff/ Logistics of the SADF. The DRDC was tasked with identifying the SADF's R&D needs and coordinating and managing R&D efforts in order to fulfil them.[85] In the same year a large part of military R&D previously carried out by the National Institute for Defence Research (NIDR) was transferred to the

---

[81] Kaplan, D., 'Ensuring technological advance in the South African manufacturing industry: some policy proposals', in Joffe *et al.* (note 66), pp. 237–63.

[82] Matthews (note 10), pp. 7–24.

[83] Legislation such as the Armaments Development and Production Act no. 57 of 1968 (as amended) prohibited the disclosure of any information relating to 'the acquisition, supply, marketing, importation, development, manufacture, maintenance, repair of, or research in connection with armaments'.

[84] United Nations Institute for Disarmament Research, *Economic Aspects of Disarmament* (UNIDIR: Geneva, 1993).

[85] Private communication with Armscor official, Pretoria, Apr. 1994.

**Table 2.13.** Military research and development expenditure, 1977–89

|  | 1977 | 1979 | 1981 | 1983 | 1985 | 1987 | 1989 |
|---|---|---|---|---|---|---|---|
| Military R&D (m. rand, constant 1985 prices) | 86 | 124 | 167 | 169 | 259 | 550 | 492 |
| Total R&D (m. rand, constant 1985 prices) | .. | 642 | 778 | 999 | 1 077 | 1 016 | 1 020 |
| Military R&D as share of total R&D (%) | .. | 19.3 | 21.4 | 16.9 | 24.0 | 54.2 | 48.2 |
| Military R&D as share of military expenditure (%) | 1.7 | 3.3 | 3.9 | 4.2 | 6.1 | 10.8 | 8.6 |
| Total R&D as share of GDP (%) | .. | 0.6 | 0.6 | 0.8 | 0.9 | 0.8 | 0.7 |

*Sources: SA Science and Technology Indicators* (Foundation for Research Development: Pretoria, 1993); and private communications with Armscor officials (value of military R&D).

Armscor subsidiary Kentron.[86] In 1984 a new management system for coordinating military R&D was established. The DRDC remained the centre of the system, tasked with identifying, funding and managing military R&D activities. It included representatives from the SADF, Armscor and the CSIR. Its interaction with other research bodies, Armscor subsidiary companies, private-sector defence companies and the universities was coordinated by the Technology Secretariat, which in turn comprised six technology work groups on vehicles, aircraft and missiles, naval vessels, electronics, weapons and ammunition, and miscellaneous programmes.[87]

After 1977 military R&D was funded from the defence budget either through the Special Defence Account or from Armscor's operating subsidy (paid out of the General Support Programme of the defence budget). The Special Defence Account, established in 1974 by an Act of Parliament and administered by Armscor on behalf of the SADF, is used to fund all major procurement projects (e.g., Cheetah fighters, Rooivalk attack helicopters and Olifant tanks), spares, parts and maintenance for SADF equipment, associated R&D, and the activities of military intelligence. Details of the Special Defence Account are not publicly available under the 1974 Special Defence Account Act, and the annual defence budget which is presented to parliament does not contain any detailed breakdown of it.[88] Military R&D activities were carried out in some of the Armscor subsidiary companies (such as Kentron and Advena), a number of state institutions (such as the CSIR), various universities and private-sector defence companies.[89]

South Africa's expenditure on military R&D increased dramatically during the 1980s. As a percentage of total R&D expenditure it increased from 19.3 per

---

[86] Landgren (note 4), p. 42.
[87] Private communication with Armscor official, Pretoria, Apr. 1994.
[88] Cawthra (note 22), p. 96.
[89] Landgren (note 4), pp. 49–53.

**Table 2.14.** Patenting activity, 1977–89

|      | Military R&D expenditure (constant m. rand, 1985 prices) | No. of S. African patents registered in South Africa | South Africa share in US patents (%)[a] |
|------|------|------|------|
| 1977 | 86   | ..    | 0.29 |
| 1978 | 98   | ..    | 0.33 |
| 1979 | 124  | ..    | 0.34 |
| 1980 | 133  | 827   | 0.30 |
| 1981 | 167  | 1 059 | 0.42 |
| 1982 | 157  | 1 436 | 0.30 |
| 1983 | 169  | 1 417 | 0.25 |
| 1984 | 178  | 1 413 | 0.28 |
| 1985 | 259  | 1 278 | 0.30 |
| 1986 | 422  | 1 162 | 0.27 |
| 1987 | 550  | 1 088 | 0.27 |
| 1988 | 515  | 1 231 | 0.28 |
| 1989 | 492  | 1 211 | 0.30 |

[a] South African patents as a percentage of all foreign patents registered in the USA.

*Sources:* Private communications with Armscor officials, Pretoria, Apr. 1994; US Patent Office, *All Technologies Report, 1990* (US Patent Office: Washington, DC, 1990); and *SA Science and Technology Indicators* (Foundation for Research Development: Pretoria, 1993).

cent in 1979 to 54 per cent in 1987, before declining to 48.2 per cent in 1989. However, the most marked increases in the late 1980s occurred during a period when total R&D expenditure and private-sector R&D were declining in real terms.[90] It seems therefore that the allocation of national resources to military R&D during the 1980s may have crowded out investment in non-military R&D.

This under-investment in civilian R&D during the 1980s corresponded with slower growth in the manufacturing sector compared with the period before 1977. It is possible that the crowding out of civilian R&D by military R&D had serious consequences for the growth, development and innovative capabilities of the civilian manufacturing sector. South Africa's total expenditure on R&D declined more rapidly than capital expenditure during the late 1980s. Furthermore, the number of scientists and engineers per capita of the population in South Africa was lower than many other developing countries (e.g., Brazil and South Korea) in the late 1980s, and the gap between South Africa and these countries increased during the late 1980s.[91]

The possible opportunity costs associated with diverting scarce R&D resources to the military sector during the 1980s were evident not only in the decline in civilian R&D expenditure during the same period but in the stagnation of technological activity as measured by scientific outputs. South Africa's patenting activity, both in South Africa and internationally, declined quite sig-

[90] Kaplan (note 81), pp. 239–41.
[91] Kaplan (note 81), pp. 239–43.

nificantly during the late 1980s (see table 2.14). This decline corresponded with the significant increases in expenditure on military R&D during the same period. This suggests that military R&D did not have a positive impact on the country's technological performance, and that spin-offs from defence to the civilian sector during the 1980s were insignificant, despite the increases in military R&D expenditure.[92]

## VI. Conclusions

This chapter describes the origins, development, size, structure and economic significance of South Africa's domestic arms industry during the apartheid era and particularly during the militarization period between 1975 and 1989. During this period the South African state and society became increasingly militarized as a result of the growing influence of the military at political, economic and social levels. The links between apartheid's internal and external dynamics, such as South Africa's involvement in a number of regional conflicts, growing internal unrest and the country's increasing international isolation, strengthened the position of the military within the state and thereby contributed to the process of militarization at all levels of society.

The militarization of the economy was reflected in increases in military expenditure, the development and expansion of the domestic arms industry and the emergence of a powerful local military–industrial complex as a result of the growing convergence of economic interests between the state, the military and the private sector with regard to defending white minority rule against the 'total onslaught'.

The arms industry's 'strategic' importance, resulting from the UN arms embargo and the SADF's increasing demand for armaments and equipment during the late 1970s and throughout the 1980s, meant that by the end of the 1980s the arms industry had become a significant consumer of national resources. However, its development and expansion between 1975 and 1989 occurred within the context of a deteriorating economic situation.

The information presented in this chapter suggests that the expansion of the domestic arms industry during the militarization period distorted the trajectory of the country's industrial development. It also imposed a number of long-term economic costs on the economy. The absorption of scarce resources (capital, labour and foreign exchange) and the crowding out of non-military public and private investment and of non-military R&D not only exacerbated many of the existing structural problems in the apartheid economy (such as shortages of skilled black labour) but also contributed to the underdevelopment, declining productivity and poor international competitiveness of the civilian economy. These costs, it is suggested, became increasingly significant during the 1970s and 1980s when the South African economy entered a period of long-term stagnation and decline.

[92] Willett and Batchelor (note 80), p. 17.

# 3. The changing strategic, political and economic environment, 1989–94

## I. Introduction

South Africa's transition to democracy, which started in the late 1980s and culminated with the holding of the country's first democratic, non-racial elections in April 1994, was prompted by the interaction of a number of strategic, political and economic developments, operating at both internal and external levels.

At a domestic level these developments included serious political violence, which made many parts of the country ungovernable; increasing domestic opposition to apartheid; and the complete failure of Total Strategy to co-opt a significant proportion of non-white South Africans into some kind of power-sharing deal with the white minority government. At an external level they included South Africa's increasing international isolation; the damaging effects of economic sanctions, particularly financial sanctions; and the end of the cold war and the demise of superpower rivalry in Southern Africa and other developing countries.[1] Another key external development was the SADF's loss of air superiority in southern Angola during 1988, which signalled the defeat of the military option as a means of guaranteeing the survival of white minority rule. These various developments, which were the key determinants of South Africa's changing strategic, political and economic environment between 1989 and 1994, prompted the South African Government and the liberation movements to opt for the path of negotiation and compromise.

This chapter describes the changing nature of South Africa's economic, strategic and political environment between 1989 and 1994, in order to provide a broader context for examining the processes of disarmament and defence industrial adjustment which are examined in detail in chapters 4 and 5.

## II. The changing strategic environment

South Africa's strategic environment changed quite dramatically after 1989. The end of the cold war, which led to the break-up of the former Soviet Union, effectively put an end to superpower rivalry in many developing countries, including Southern Africa. The cessation of East–West confrontation was accompanied by a reduction in ideological tensions within and between African countries and by significant moves towards political pluralism in Southern

---

[1] Ohlson, T., 'South Africa: from apartheid to multi-party democracy', *SIPRI Yearbook 1995: Armaments, Disarmament and International Security* (Oxford University Press: Oxford, 1995), p. 124.

Africa (as in Malawi and Zambia).[2] Within South Africa the end of the cold war also released the ideological stranglehold that anti-communism had on the apartheid state.

The loss of South African air superiority at Cuito Cuanavale in southern Angola in 1988 signalled the defeat of military aggression and regional destabilization as an option for defending white minority rule and prompted South Africa to reassess its military involvement in the Angolan war and its covert support to UNITA (the National Union for the Total Independence of Angola) in Angola and RENAMO (the Mozambique National Resistance Movement) in Mozambique.[3] After a series of negotiations between South Africa, Angola and Cuba, South Africa and Cuba agreed to withdraw their armed forces from Angola and Namibia, thus paving the way for Namibia's independence in March 1990. These developments prompted the MPLA (the Popular Movement for the Liberation of Angola) and UNITA to begin negotiating an end to the nearly 20-year-old civil war in Angola, and a peace agreement (the Bicesse Accords) was signed between them in 1991 which provided for the holding of democratic elections under the supervision of the United Nations. However, the civil war started again in late 1992 after UNITA rejected the results of the UN-supervised elections. The United Nations Angola Verification Mission (UNAVEM) remained in Angola despite the resumption of the civil war, and in November 1994 the two sides signed the Lusaka Agreement, signalling an end to the civil war.

The positive political developments which occurred in Angola, Namibia and South Africa, together with the end of the cold war, also created the conditions for the resolution of Mozambique's long-running civil war. In late 1992 FRELIMO (the Front for the Liberation of Mozambique) and RENAMO signed the Rome Peace Agreement, thereby ending the civil war, and in 1993 a United Nations peacekeeping force, the United Nations Operation in Mozambique (UNOMOZ), arrived to supervise the holding of democratic elections, which were scheduled for late 1993. However, because of a number of delays relating to the implementation of various aspects of the peace agreement, the elections were delayed and only took place in October 1994.[4]

Between 1991 and 1994 eight countries in Southern Africa[5] initiated political reforms, including the legalization of opposition parties, and made moves towards multi-party democratic systems. By the time of the elections in South Africa in April 1994, only three countries remained at odds with this process of democratization: Angola, which was once again immersed in civil war, Swaziland, which was still ruled by an autocratic monarchy, and Zimbabwe, which

[2] Nathan, L., 'Towards a post-apartheid threat analysis', *Strategic Review for Southern Africa*, vol. 15, no. 1 (1993), pp. 43–71.

[3] For details of South Africa's support for RENAMO, see Minter, B., *Apartheid's Contras* (Zed Books: London, 1994).

[4] United Nations, *The United Nations and Mozambique 1992–1995* (UN: New York, 1995).

[5] Angola, Lesotho, Malawi, Mozambique, Namibia, South Africa, Tanzania and Zambia. Botswana was the only multi-party democracy in Southern Africa before 1991.

embodied the characteristics of a de facto one-party state.[6] These positive domestic political developments were also accompanied by the implementation of various disarmament measures in many Southern African countries, partly in response to the ending of the cold war, the demise of apartheid, the improvement in interstate relations and growing pressure from international donor agencies. For example, between 1991 and 1994 full-time force levels were reduced from 58 000 to 11 000 in Mozambique, from 70 000 to 8100 in Namibia and from 54 600 to 46 900 in Zimbabwe.[7] Military expenditure levels also declined significantly in many countries in the region between 1990 and 1993—in Botswana from $156 million to $132 million (1990 prices); in Malawi from $24 million to $15 million; in Zambia from $146 million to $48 million; and in Zimbabwe from $390 million to $368 million.[8]

These interlinked processes of democratization and disarmament in many countries in the region had a positive impact on the South African state's external threat perceptions, and were reflected in statements made by various government ministers. In 1990, General Magnus Malan, South Africa's Defence Minister, stated in parliament:

Things have already changed dramatically in Southern Africa . . . whereas the [aim of Soviet interests] previously was to achieve a take-over of power in Southern Africa, there has been a dramatic shift in emphasis . . . that tremendous risk of pro-Communist forces has already changed dramatically . . . the conflicts that have received attention up to now in terms of East–West conflict could lose their importance completely . . . what is clear is that we need not expect an East–West conflict in the coming years in Africa . . . we need not expect an attack on South Africa out of Africa by a combined force either.[9]

The dramatic developments in South Africa's strategic environment together with internal political developments prompted significant changes in the country's defence and foreign policies. South Africa honoured its commitment to the implementation of UN Security Council Resolution 435 and withdrew its armed forces from Angola and Namibia in 1989. It formally abandoned its policy of military aggression and regional destabilization[10] and embarked on an ambitious programme of diplomatic and economic outreach to African states.[11] These

[6] Ohlson, T., Stedman, S. and Davies, R., *The New is Not Yet Born: Conflict Resolution in Southern Africa* (Brookings Institution: Washington, DC, 1994).

[7] Figures drawn from International Institute for Strategic Studies, *The Military Balance* (Brasseys: London, various issues).

[8] George, P. *et al.*, 'World military expenditure', *SIPRI Yearbook 1995* (note 1), pp. 443–44; and George, P. *et al.*, 'Military expenditure', *SIPRI Yearbook 1997: World Armaments and Disarmament* (Oxford University Press: Oxford, 1997), pp. 198–99.

[9] South African House of Assembly, *Debates*, 17 May 1990, cols 9529–33.

[10] Despite its formally abandoning its policy of support to UNITA and RENAMO, evidence has emerged that South Africa, or elements within the South African state, continued to provide covert support to RENAMO up to the signing of the Rome Agreement in 1992 and to UNITA up to the elections in South Africa in Apr. 1994. Minter (note 3).

[11] Nathan, L. and Phillips, M., 'Cross-currents: security developments under F. W. de Klerk', eds G. Moss and I. Obery, *South African Review 6: From 'Red Friday' to CODESA* (Ravan Press: Johannesburg, 1992).

developments, together with the start of constitutional negotiations aimed at ending apartheid in mid-1990, removed the dominant source of instability and antagonism in the region and led to a dramatic improvement in relations between South Africa and the other countries of Southern Africa. Concrete examples of this improved regional security environment were Zimbabwe's cancellation of an order for MiG-29 fighter aircraft in 1991[12] and the withdrawal of Malawian and Zimbabwean troops from Mozambique in late 1992 as a result of the peace agreement between FRELIMO and RENAMO.

The visible improvement in relations between South Africa and the rest of the region created the conditions for greater regional cooperation on defence and security matters. In 1992 as the transition to democracy in South Africa became increasingly irreversible the Southern African Development Co-ordination Conference (SADCC), which had been set up by the countries in Southern Africa to reduce their economic dependence on South Africa, transformed itself with the Windhoek Declaration[13] into the SADC and committed itself to promoting greater regional integration and cooperation on a number of issues, including defence and security.[14]

This commitment to greater regional cooperation on defence and security issues was influenced by the proposal at a meeting in Kampala, Uganda in 1991 to establish a Conference on Security, Stability, Development and Co-operation in Africa (CSSDCA), modelled on the Conference on Security and Co-operation in Europe (CSCE).[15] The CSSDCA proposal stressed the critical relationship between peace and development on the African continent, as well as emphasizing that the security and stability of each African country were inescapably connected to the security of surrounding states. It also adopted an integrated approach to security, and grouped proposals in four 'calabashes': security, stability, development and cooperation.[16]

After its inception in 1992 the SADC pursued the issue of regional cooperation on defence and security matters at the level of political rhetoric. The Windhoek Declaration stated that: 'a new Southern Africa, concerned with peace and development, must find a more abiding basis for continuing political solidarity and cooperation in order to guarantee mutual peace and security in the region'.[17] The 1993 the SADC Framework and Strategy for Building the Community proposed the following strategies for enhancing regional security: the adoption of a 'new approach to security' which emphasized the security of

---

[12] *Jane's Defence Weekly*, 20 July 1991, p. 115.

[13] Southern African Development Community, Towards the Southern African Development Community: A Declaration by the Heads of State or Government of Southern African State, SADC, Windhoek, 1992, p. 5.

[14] Nathan, L. and Honwana, J., *After the Storm: Common Security and Conflict Resolution in Southern Africa*, Arusha Papers no. 3 (Centre for Southern African Studies, University of the Western Cape: Bellville, Feb. 1995).

[15] Nathan, L., 'Beyond arms and armed forces: a new approach to security', *South African Defence Review*, no. 4 (1992), pp. 12–21.

[16] Africa Leadership Forum, Towards a Conference on Security, Stability, Development and Co-operation in Africa [the Kampala Document], Africa Leadership Forum: Kampala, 1991.

[17] Towards the Southern African Development Community (note 13), p. 5.

people and the non-military dimensions of security; the creation of a forum for mediation and arbitration; reductions in force levels and military expenditure; the introduction of confidence- and security-building measures and non-offensive defence doctrines; and the ratification of key principles of international law governing interstate relations.[18] Notwithstanding these pronouncements, the institutional arrangements within which regional cooperation on defence and security matters would take place only began to take shape after the elections in South Africa in April 1994 and after South Africa had officially joined the SADC in August 1994.

South Africa's formal admittance into the SADC signalled not only that its rehabilitation into the region and the continent was complete, but that future defence and security issues in the region would be dealt with in a collective and collaborative manner. Furthermore, the fact that South Africa's strategic environment became increasingly benign between 1989 and 1994 supported the process of democratization which was occurring in South Africa and provided added impetus to the process of disarmament which accompanied South Africa's transition to democracy.

## III. The changing political environment

The dramatic changes in South Africa's strategic environment after 1989, largely as a result of the end of the cold war, had a profound effect on the domestic political setting. The causal relationship between changes in the external strategic environment and domestic political developments was the result of the fact that since its inception apartheid had acquired interrelated internal and external dynamics. For example, the official threat perception of Soviet expansionism in Southern Africa was used by the South African regime to gain tacit or indirect support from the West, dampen anti-apartheid sentiments in Europe and the USA and mobilize the white minority at home. Thus, despite being an internal socio-political policy, apartheid became increasingly internationalized through South Africa's acts of military aggression and regional destabilization, and through the international community's growing opposition to apartheid.

South Africa's military withdrawal from Angola and Namibia had a number of positive effects on the domestic political scene, while positive internal political developments—the start of constitutional negotiations aimed at ending apartheid—had a significant impact on its relations with the region and the rest of the world. The recognition that the military option was no longer a viable means of guaranteeing the long-term survival of white minority rule forced the de Klerk Government to pursue negotiation and accommodation as the only solutions to South Africa's internal political problems. At the same time the ANC and the various liberation movements had run out of alternatives, given that the overthrow of the apartheid state through guerrilla warfare was never a

[18] Southern African Development Community, Southern Africa: A Framework and Strategy for Building the Community, SADC, Harare, 1993, pp. 24–26.

credible option.[19] In the context of this stalemate both the South African Government and the liberation movements realized that neither side could defeat the other and that the choice was no longer between victory and capitulation but between negotiation and disaster.[20]

The election of F. W. de Klerk as President in 1989 marked the beginning of a fundamental change in South Africa's domestic political environment. At a domestic level it had became increasingly clear to the de Klerk Government by the end of the 1980s that the previous government's security and counter-revolutionary policy known as Total Strategy, aimed at defending and maintaining white minority rule, was neither viable nor appropriate. According to one commentator, '"Total Strategy" [had] failed to create a new supportive political alliance capable of resolving apartheid's legitimacy crisis . . . . [and] under crisis-imposed strain there was gradual weakening of white political cohesion regarding the path South Africa ought to take'.[21] In this context the de Klerk Government concluded that the mixture of limited reform and intensified repression which had characterized much of the later years of Total Strategy would not solve the country's pressing domestic political problems and that fundamental reforms were necessary. From late 1989 de Klerk initiated a series of political reforms. In early 1990 he unbanned all opposition parties, including the ANC, released political prisoners (such as Nelson Mandela), lifted the national state of emergency imposed in 1986, granted indemnity to hundreds of anti-apartheid activists in exile and relaxed many of the prohibitions against political activities such as public demonstrations and marches.[22] In 1991 he also abolished the remaining legislative cornerstones of apartheid and amended certain aspects of security legislation.

These reforms led to the start of constitutional negotiations between the government and the ANC aimed at ending apartheid and were instrumental in altering the pattern of civil–military relations which had existed during the apartheid era. They were accompanied by the liberalization of state security policies and the implementation of a number of disarmament measures, which had a profound impact on the size and position of the country's arms industry. The constitutional negotiations started with the Convention for a Democratic South Africa (CODESA I) in late 1991 and continued until the establishment of the Transitional Executive Council (TEC) in December 1993.[23] The negotiations (CODESA I and II and the Multi-party Negotiating Forum) were characterized by a number of breaks as the various parties attempted to reach agreement on various issues.[24] However, by the end of 1993 a great degree of

[19] Ohlson (note 1), p. 124.

[20] Maphai, V., 'The politics of transition to 1990', ed. V. Maphai, *South Africa: The Challenge of Change* (SAPES Books: Harare, 1994), p. 64.

[21] Ohlson (note 1), p. 124.

[22] Nathan and Phillips (note 11), pp. 114–16.

[23] For a discussion of de Klerk's reforms, see Friedman, S. and Atkinson, D. (eds), *South African Review 7: The Small Miracle. South Africa's Negotiated Settlement* (Ravan Press: Johannesburg, 1994), pp. 8–10.

[24] Friedman, S. (ed.), *The Long Journey: South Africa's Quest for a Negotiated Settlement* (Ravan Press: Johannesburg, 1993).

consensus had been reached on a number of issues, including a date—27 April 1994—for South Africa's first non-racial democratic elections.

The TEC Bill was sent to parliament on 9 September after the negotiators had accepted it by 'sufficient consensus', and became law on 9 November 1993. The establishment of the multi-party TEC in December 1993 reflected a compromise in that the de Klerk Government stayed in power, but only as an interim body.[25] The purpose of the TEC was to 'level the political playing-field' and promote a climate in which people could participate freely in politics and in free and fair elections. Seven sub-councils, on defence, law and order, intelligence, regional and local authorities, finance, foreign affairs and the status of women, were established under the supervision and control of the TEC. Once the TEC Bill was approved, constitutional negotiators began the task of finalizing the interim constitution. That constitution, which some commentators have called an élite settlement, was completed on 2 December 1993 and sent to parliament for approval.[26] The next week the TEC took office, and on 18 December 1993 the interim constitution, Act no. 200 of 1993, was passed.[27]

A dominant feature of de Klerk's reforms after 1989 was the re-establishment of civilian control over the military and the liberalization of certain aspects of state security policy. According to Williams, de Klerk, unlike his predecessor, P. W. Botha, did not have a political power base within the security establishment; rather his power bases were civilian in composition and ethos and resided in departments such as Constitutional Planning and Foreign Affairs.[28] Unlike that of the 'securocrats' of the Botha Administration, de Klerk's strategy lacked the managerial overtones of the technocratic-reformist approach, and was premised more on consensus and incorporation.[29]

In terms of concrete steps aimed at reclaiming civilian control of the state, de Klerk ensured that the State President's Office and the cabinet replaced the State Security Council (SSC) as the locus of state power. The SSC remained as a statutory body, but a new Cabinet Committee for Security Affairs (CCSA) was established as one of four standing cabinet committees and its decisions were made subject to cabinet endorsement.[30]

The National Security Management System (NSMS) was replaced by a National Coordinating Mechanism (NCM) in 1990 in order to alter the perception that Total Strategy-oriented securocrats were still dominating policy making and to fit in with the government's new reform initiatives.[31] The NCM was

[25] Sarakinsky, I., 'Rehearsing joint rule: the Transitional Executive Council', eds Friedman and Atkinson (note 23), p. 68.

[26] Atkinson, D., 'Brokering a miracle? The Multi-party Negotiating Forum', eds Friedman and Atkinson (note 23), p. 35.

[27] Constitution of the Republic of South Africa, Act no. 200 of 1993, *Government Gazette*, no. 15466, 28 Jan. 1994.

[28] Williams, R., *The Changing Parameters of South African Civil–Military Relations: Past, Present and Future Scenarios*, Working Paper no. 6 (Military Research Group: Johannesburg, 1993), p. 9.

[29] Williams (note 28), p. 9.

[30] Seegers, A., 'Current trends in South Africa's security establishment', *Armed Forces and Society*, vol. 18, no. 2 (winter 1992), pp. 159–74.

[31] Hough, M. and du Plessis, A., *Selected Official South African Strategic Perspectives: 1989–1992*, ISSUP Ad Hoc Publication no. 29 (Institute for Strategic Studies, University of Pretoria: Pretoria, 1992),

intended to ensure coordination between the cabinet, the four cabinet commit-tees (of which the committee for security affairs was only one), top government officials and departmental task groups. Initially the NCM came under the con-trol of the State President's Office, but in November 1991 it was placed under the control of the Department of Regional and Land Affairs.[32]

De Klerk also initiated fundamental changes in the intelligence services in order to ensure civilian dominance in intelligence matters. The National Intelli-gence Service (NIS), which was linked to the Department of Foreign Affairs, replaced the Directorate of Military Intelligence and the security police as the central intelligence agency within the state and took control of the SSC secre-tariat away from the SADF and the South African Police.[33] The NIS was thus restored to the apex of the intelligence pyramid and was entrusted with the tasks of overall intelligence coordination and the formulation of the strategic intelli-gence brief. However, its relatively small size in terms of personnel meant that its capacity to control the other parts of the security establishment was exceed-ingly limited.[34]

Other measures were also introduced to support civilian control over the military. Civilian politicians such as Roelf Meyer and Kobie Coetzee were appointed as defence ministers in preference to soldiers such as General Malan, who had been Defence Minister under P. W. Botha, in an attempt to diminish the political influence of the SADF within the cabinet. De Klerk also excluded SADF and Ministry of Defence personnel from his negotiating teams on CODESA I and II and took steps to undercut the status and influence of the military in a number of ways.[35]

Conscription for white males was reduced from two years to one year in 1991 and abolished at the end of 1993. The defence budget was cut by over 40 per cent in real terms between 1989 and 1993, most of the cuts coming from reduc-tions in procurement expenditure.[36] South Africa's nuclear weapon programme was terminated by de Klerk in 1989 and South Africa signed the 1968 Non-Proliferation Treaty (NPT) in July 1991.[37] In the following years South Africa supported a number of multilateral disarmament initiatives, particularly with respect to weapons of mass destruction.[38]

De Klerk's attempts to 'demilitarize' the state and reduce the power and influence of the military establishment were characterized by a number of

p. 4. See also Seegers, A., 'South Africa's national security management system: 1972–1990', *Journal of Modern African Studies*, vol. 29, no. 2 (1991), pp. 253–73 for a discussion of the NSMS during the 1970s and 1980s.

[32] Hough and du Plessis (note 31), p. 6.
[33] Nathan and Phillips (note 11), p. 115.
[34] Williams (note 28), p. 12.
[35] Nathan and Phillips (note 11), p. 115.
[36] Batchelor, P., 'Conversion of the South African defence industry: prospects and problems', Paper presented at Military Research Group workshop on Arms Trade and Arms Conversion in a Democratic South Africa, Pretoria, 28–30 June, 1993.
[37] Hough, M., 'Disarmament and arms control with specific reference to the RSA', *South African Defence Review*, no. 9 (1993), pp. 17–23.
[38] Cilliers, J., 'Towards a South African conventional arms trade policy', *African Security Review*, vol. 4, no. 4 (1995), p. 6.

inconsistencies and remained incomplete by the time of the elections in April 1994.[39] While at a formal level he implemented a number of political reforms and disarmament measures which were designed to reduce the influence of the military, he failed to take steps which would have fundamentally challenged its institutional and ideological interests. The government's apparent inconsistencies with regard to the military were based on a decision taken at a cabinet meeting in early 1990 that the SADF should remain untouched as the 'stable core around which the dynamic of change would occur' and that the government alone would control the military until it was reasonably certain of the outcome of the transition.[40]

The fact that increasing levels of political violence accompanied the progress of the constitutional negotiations aimed at ending apartheid led to suspicions that there were elements in the state who were intent on sabotaging the negotiation process. Concrete evidence began to emerge in 1991 that a 'Third Force' within the security forces was responsible for much of the violence.[41] De Klerk made some attempts to curb the alleged covert actions of the security forces through the establishment of judicial commissions of enquiry (including the Goldstone Commission) and an internal SADF inquiry led by Lieutenant-General Pierre Steyn which led to the subsequent purges of 23 senior SADF officers suspected of Third Force activity.[42] However, none of the officers was charged, and all received substantial severance packages. As with the purge of senior police officers, the military's attempt to break with its past was less total than it seemed: neither General George Meiring (then chief of the army) nor the Chief of Staff of Military Intelligence, Joffel van der Westhuisen, lost his job, although they were also allegedly indirectly implicated by the Goldstone Commission.[43]

Recent revelations have confirmed the thesis that the de Klerk Government pursued a double-edged political agenda after the unbanning of the liberation movements in 1990. While engaged in formal constitutional negotiations on the one hand, on the other hand it sought to destabilize the ANC through orchestrating violence to coincide with major ANC activities and through targeting ANC leaders for vilification and assassination campaigns.[44] As a result of this double-edged agenda, and because de Klerk lacked any meaningful influence or operational control over the SADF, he was unwilling and unable to control or fully rein in the 'dissident elements' within the security forces who were accused of

[39] Nathan and Phillips (note 11).

[40] Shaw, M., 'Biting the bullet: negotiating democracy's defence', eds Friedman and Atkinson (note 23), pp. 228–56.

[41] While the notion of a Third Force is ambiguous, it is generally agreed that the origin of Third Force-related activities is to be found in the military (the Department of Military Intelligence and the Special Forces), the police and various 'hit squads' linked to elements of the security forces. Ellis, G., *Third Force: What Evidence?*, Regional Topic Paper, vol. 93, no. 1 (South African Institute of Race Relations: Johannesburg, May 1993).

[42] Shaw (note 40), p. 229.

[43] Shaw (note 40), p. 229.

[44] 'Former agent details destabilisation campaign', *Southscan*, vol. 10, no. 26 (7 July 1995), pp. 201–202.

fomenting or perpetuating political violence through acts of commission or omission.[45] The government's failure in this regard forced it into an embarrassing admission of deceit in its negotiation strategies, which in turn undermined its position *vis-à-vis* the ANC in the political negotiations.[46]

Despite the political reforms and the start of talks between the government and the ANC in mid-1990, defence matters such as the issues of integrating armies and establishing a post-apartheid defence force were not formally discussed as part of the constitutional negotiations until late 1993.[47] The time-lag between the political negotiations and the start of formal negotiations on defence issues was related to the fact that both the ANC and the government saw the retention of a 'security fall-back' as a critical means of appeasing their constituents, who were still wary of the compromises being brokered in the political negotiations. The security fall-back was also seen as a physical guarantee should the political negotiations falter or fall apart.[48]

The process of formal negotiations between the government and the ANC on military matters was initiated by the military, not the politicians, and from early 1993 many SADF generals pressed hard for formal discussions with Umkonto we Sizwe (MK).[49] The SADF's desire to negotiate with the ANC was based on the perception that it could not remain completely isolated from the political changes that were occurring and that if it entered into talks with the ANC (and the MK) then it would be better able to manage change (and pre-empt ANC demands). The first formal meeting between the SADF (representatives of the air force) and the ANC took place in March 1993 over the proposed purchase of trainer aircraft from Switzerland. Members of the MK were present. While the matter was not resolved, it laid the foundations for the first formal meeting between the SADF and the MK, which was held in Simonstown in April 1993. This meeting discussed a number of military-related issues and established specialist working groups to examine three key sets of issues: (a) control of the armed forces during the transitional phase before elections; (b) the establishment of a national peacekeeping force; and (c) the integration of all armed forces into a new defence force after the elections.[50] The working groups, which included representatives from the SADF and MK, produced a number of proposals over the next few months, including the clauses affecting defence

[45] Nathan and Phillips (note 11), pp. 112–13.

[46] Nathan and Phillips (note 11), p. 123.

[47] Informal discussions between the SADF and Umkhonto we Sizwe (MK) started in 1990 at the Lusaka Conference and continued until the start of formal negotiations in late 1993. SADF and MK sources confirm that between 6 and 8 informal meetings took place before the first official meeting in Mar. 1993. Nathan, L., 'Riding the tiger: the integration of armed forces and post-apartheid military', *Southern African Perspectives*, no. 10 (Centre for Southern African Studies: Bellville, 1991); and Shaw (note 40), p. 238.

[48] Williams, R., *South Africa's New Defence Force: Progress and Prospects*, CSIS Africa Notes no. 170 (Centre for Strategic and International Studies: Washington, DC, Mar. 1995), p. 2.

[49] SADF Chief of Staff Pierre Steyn, who led the internal investigation into the conduct of senior SADF officers in 1992, argued during 1992 that the SADF should initiate talks with MK in order to gain greater control over the way in which change, both political and within the SADF, would be managed. Shaw (note 40), pp. 230–31.

[50] Shaw (note 40), p. 240.

which would be incorporated into the interim constitution. The significance of the meeting lay in the fact that the SADF had to concede that, despite its greater military strength, its future role in the post-apartheid era was dependent upon the MK's legitimacy.

The TEC Sub-Council on Defence, created in December 1993, was assigned responsibility for oversight of the armed forces during the run-up to the elections and for initiating planning relating to the creation of a new defence force. Its role was essentially political and strategic. It was also empowered to establish and maintain a national peacekeeping force.[51]

The negotiations on the defence provisions of the interim constitution began immediately after the TEC Act became law on 9 November 1993. The ANC's policy on defence matters was drawn largely from the work of the Military Research Group (MRG), a loose grouping of analysts who functioned as an unofficial policy-making group on defence and security matters for the ANC.[52] The central thrust of the ANC's position was ensuring the establishment of democratic civil–military relations, that is, civil supremacy over the armed forces, and the accountability of the armed forces to the public through parliament. The ANC also adopted a relatively anti-militarist position, largely because of the influence of the MRG, and argued that the 'defence force shall be defensive in its character, orientation and its strategy, and its force levels will be adjusted accordingly' and that 'South Africa shall be committed to resolving conflicts primarily through non-violent means'.[53]

The SADF's key negotiating principles were the 'retention of standards' and the 'apolitical nature and character of the defence force', for they were more intent on protecting the military than on separating it from political affairs.[54] The SADF negotiating team also had the following objectives: that the SADF should form the core of the new defence force; that SADF structures should be retained; that the future of all SADF members must be assured; and that other forces should be integrated into existing SADF structures.[55] The critical debate over integration, which the ANC favoured, or absorption, which the SADF favoured, was won by the ANC insofar as the name of the SADF was changed to the South African National Defence Force (SANDF). However, the size, power and managerial capacity of the SADF meant that, in practice, it would absorb other armed formations.

Another key negotiating issue was political control of the military. The SADF wanted the new president to exercise powers over the military only with the concurrence of the deputy presidents and the cabinet, while the ANC position, which was accepted in the interim constitution, was that the military could only

[51] *Race Relations Survey 1993/94* (South African Institute for Race Relations: Johannesburg, 1994), pp. 506–507.

[52] Shaw (note 40), p. 233.

[53] 'ANC policy guidelines', ANC National Conference, 28–31 May 1992, section Q(3), pp. 46–47.

[54] Shaw (note 40), p. 233.

[55] 'Achievements by the SADF in the negotiating process', *CSADF Internal Communication Bulletin*, no. 2 (Jan. 1994).

be deployed subject to the authority of parliament.[56] Both parties, however, agreed to the establishment of a joint standing committee of parliament on defence, consisting of members from all parties in the national assembly, with the authority to investigate and make recommendations regarding the budget, functioning, organization, armaments, policy, morale and state of readiness of the defence force.

The SADF also used the constitutional negotiations to emphasize its concern with the maintenance of standards. Thus it argued for provisions in the constitution such as 'the SANDF shall be established in such a manner that it will provide a balanced, modern and technologically advanced military force' and 'all members of the National Defence Force shall be properly trained to comply with international standards of competency'.[57] These provisions, according to some commentators, were not only intended to halt the cuts in defence expenditure, which had curtailed expenditure on new equipment and resulted in the down-sizing of the domestic arms industry, but were a covert attempt by whites within the SADF to hold on to military power, given the fact that most of the MK cadres were infantry soldiers and not trained in the use of highly sophisticated weaponry.[58] The provision for 'a balanced, modern and technologically advanced military force', which the SADF insisted on, was balanced by the provision 'the national defence force shall be primarily defensive in the exercise or performance of its powers and functions', which was drawn directly from the ANC's Policy Guidelines.[59] By early December the negotiations on the defence provisions in the interim constitution were completed, and the ANC and SADF negotiators turned their attention to the mechanics of integration, and in particular the issue of assembly points.

Operational and tactical responsibility for strategic planning for the integration process and the creation of a new defence force was delegated by the Sub-Council on Defence to the Joint Military Co-ordinating Council (JMCC), which was established in January 1994. The JMCC was co-chaired by the SADF and the MK and consisted of representatives of all armed forces with political representation in the TEC—the SADF, the MK, and the armies of the homelands of Transkei, Bophutatswana, Venda and Ciskei (collectively referred to as the TBVC armies). It was divided into a number of working groups according to functional area (e.g., logistics and personnel), arm of service (e.g., army and navy), or specific issues for consideration (e.g., uniforms).[60] As a result of the establishment of the JMCC, in practice the Sub-Council on Defence functioned as a structure within which negotiations on the military could take place.[61]

The JMCC planning process, which took place from January 1994 until just before the elections in April 1994, reached agreement on two important issues, namely, (a) force design, including the rationalization of certain structures and

[56] Shaw (note 40), p. 234.
[57] Constitution of the Republic of South Africa (note 27), sections 226 (4) and 226 (5).
[58] Shaw (note 40), p. 234.
[59] Constitution of the Republic of South Africa (note 27), sections 226 (4) and 227 (2)(f).
[60] Williams (note 48), p. 2.
[61] Shaw (note 40), pp. 241–42.

the renaming of certain units, and (*b*) the principles, procedures and mechanisms for integration, including the identification of assembly areas and the involvement of a British Military Advisory Training Team (BMATT) to oversee the integration process. During the planning for integration it was anticipated that the MK would deliver some 20 000 personnel, the four TBVC armies around 11 000 and the former SADF some 90 000. Thus it was estimated that the new SANDF would have an initial strength of around 120 000 when it came into being on 27 April 1994.

The functioning of the Sub-Council on Defence, and particularly the JMCC planning process, was dominated by the SADF and the MK, while the TBVC armies played a relatively marginal role.[62] The SADF exerted a disproportionate influence over the entire strategic planning process, given its virtual monopoly of formal staff skills and strategic management concepts, its keen sense of bureaucratic politics, and its familiarity with the practical, conceptual, strategic and doctrinal issues underpinning military planning and force design.[63] However, the MK's sense of popular legitimacy, which the SADF did not possess, and the fact that it was able to exercise an important veto power over the entire process, since no agreement could be reached without its participation, prevented strategic planning from being completely dominated by the SADF.

The issue of the future of the domestic arms industry was not central to the role either of the Sub-Council for Defence or of the JMCC. However, the TEC Bill gave the Sub-Council on Defence the authority 'to undertake or to commission research into . . . the future of the arms and related industries'.[64] In March 1994 the Sub-Council appointed a working group to draw up a draft national policy for the arms industry. This small, self-selecting working group, which was dominated by representatives from the SADF and Armscor, presented its findings in the form of a draft document entitled 'National Policy for the Defence Industry' to the Sub-Council on Defence in April 1994. The document argued that the arms industry represented 'a substantial national asset' and should be retained subject to the following considerations: that it should be appropriately sized for national and regional needs and that it should be allowed to export its products in a responsible way while simultaneously making an important contribution to industrial development by employing its technology and capabilities in the civilian economy.[65] While the document made some vague concessions to the notion of encouraging diversification and conversion, its main thrust was towards gaining political support for preserving the defence industrial base in its present form, thereby ensuring that those with vested inter-

---

[62] The fact that the SADF began to prepare its negotiating positions in late 1992, without the direct knowledge of the politicians and months before the start of formal negotiations with the MK, gave them an added advantage in the JMCC process. Shaw (note 40), p. 237.

[63] Williams (note 48), pp. 2–3.

[64] Transitional Executive Council Bill, no. 162 of 1993, section 16 (2)(f) (Government Printer: Pretoria, 1993).

[65] Transitional Executive Council Sub-Council on Defence, National Policy for the Defence Industry (TEC Sub-Council on Defence: Pretoria, Apr. 1994), p. 3.

ests in the arms industry would retain the power and influence which they had exercised during the apartheid era.

South Africa's domestic political environment thus saw momentous changes between 1989 and April 1994 as the country moved towards the establishment of a democratic political order. Dominant features of this change were the re-establishment of civilian control over the military and the attempts by de Klerk to demilitarize the state and limit and undermine the influence and status of the military through budget cuts and various disarmament measures. However, as a result of the government's double-edged political agenda, the process of disarmament during this period was not accompanied by a thorough transformation of the military establishment or by a broader process of demilitarization at political, institutional and ideological levels. The SADF's dominance of the negotiations on defence issues also ensured that the institutional and ideological manifestations of militarism from the apartheid era remained largely intact and unchallenged during the transition period and into the post-election era.

## IV. The changing economic environment

The political economy of apartheid had acted as a constraining factor on South Africa's socio-economic development, exacerbating internal tensions and contributing to the delegitimization of apartheid even in the eyes of many who had previously supported the regime. The growing crisis of the economy between 1989 and 1993 forced de Klerk to introduce macroeconomic reforms that were to fundamentally erode the privileged resource base of the military–industrial complex. This section provides a brief overview of the growing macroeconomic constraints that were to force the budget-led disarmament measures which are discussed in chapter 4.

The severe deterioration in South Africa's economic situation between 1989 and 1994 certainly played a significant role in prompting de Klerk to initiate his series of political reforms during the same period. It also at some points undermined the positive progress that was being achieved on the political front, by exacerbating the country's myriad economic problems (such as high unemployment) inherited from the apartheid era.

When de Klerk came to power in 1989 he was confronted with a deteriorating economy, largely as a result of trade and financial sanctions and South Africa's increasing international economic isolation. The economic crisis also had its roots in the economic system created by apartheid, which was characterized by great disparities between whites and blacks in terms of wealth, income, health, education, housing and land.[66] The dysfunctional nature of the apartheid economic system became increasingly obvious during the 1980s with declining levels of GDP and fixed capital formation, massive unemployment, growing income inequality between blacks and whites, substantial capital flight and the

---

[66] Ohlson (note 1), p. 121.

**Table 3.1.** Major economic indicators, 1989–94

Figures are percentages.

| | 1989 | 1990 | 1991 | 1992 | 1993 | 1994 |
|---|---|---|---|---|---|---|
| GDP growth | 2.4 | – 0.3 | – 1.0 | – 2.2 | 1.3 | 2.8 |
| GDP per capita growth | 0 | – 2.6 | – 3.3 | – 4.3 | – 0.9 | 0.5 |
| Inflation | 14.7 | 14.4 | 15.3 | 13.9 | 9.7 | 9.0 |
| Growth in employment[a] | . . | – 0.3 | – 1.7 | – 2.0 | – 2.1 | – 0.6 |
| GDFI[b] growth | 6.5 | – 2.3 | – 7.4 | – 5.3 | – 2.8 | 8.7 |
| GDFI as share of GDP | 20.6 | 19.6 | 17.8 | 16.6 | 15.5 | 16.0 |

[a] Growth in total non-agricultural employment.

[b] Gross domestic fixed investment.

*Source:* South African Reserve Bank, *Quarterly Bulletin* (South African Reserve Bank: Pretoria, various issues).

increasing structural problems associated with maintaining the apartheid economy, including the high levels of military expenditure required to maintain the military–industrial complex.[67] The recession—the longest and worst since the great depression of the 1930s—was influenced by the global recession, particularly among South Africa's major trading partners; the government's inability to control expenditure, which aggravated domestic inflation; a severe drought in 1992–93 which made food imports necessary; high levels of domestic political violence which accompanied the transition to democracy; and uncertainty surrounding the possible implementation of socialist economic policies by a future ANC Government. As a result of the deteriorating economic environment, a number of meetings between the ANC and the white business community were held between 1985 and 1989.[68]

South Africa experienced negative economic growth and declining GDP per capita for most of the period between 1989 and 1993 (see table 3.1). Although inflation was slowly brought under control during the early 1990s as a result of tight Reserve Bank monetary policies (including the maintenance of positive real interest rates), this was achieved at the expense of economic growth and employment, which in turn had a negative effect on the general socio-political environment. One of the most dramatic features of this period was the significant decline in gross domestic fixed investment, both in real terms and as a percentage of GDP.[69] Total employment in the non-agricultural sector declined. However, employment in the government sector continued to grow during this period and in 1993 accounted for 30 per cent of total non-agricultural employment, up from 28 per cent in 1989.

[67] Gelb, S. (ed.), *South Africa's Economic Crisis* (David Philip: Cape Town, 1991); Schrire, R. (ed.), *Critical Choices for South Africa: An Agenda for the 1990s* (Oxford University Press: Cape Town, 1991); and Simkins, C., 'The South African economy: problems and prospects', ed. J. Spence, *Change in South Africa* (Pinter: London, 1994), p. 65.

[68] Ohlson (note 1), p. 120.

[69] Joffe, A., Kaplinsky, R., Kaplan, D. and Lewis, D., *Improving Manufacturing Performance in South Africa* (University of Cape Town Press: Cape Town, 1995), p. 10.

**Table 3.2.** Public finance, 1989–94

|  | 1989 | 1990 | 1991 | 1992 | 1993 | 1994 |
|---|---|---|---|---|---|---|
| Government expenditure | 87.9 | 81.4 | 82.5 | 90.3 | 96.7 | 88.4 |
| (b. rand, constant 1990 prices) | | | | | | |
| % change | | – 7.4 | 1.4 | 9.5 | 7.1 | – 8.6 |
| as share of GDP (%) | 31.7 | 29.4 | 30.1 | 33.7 | 35.7 | 31.7 |
| Government revenue | 75.4 | 72.1 | 68.7 | 65.0 | 68.4 | 72.4 |
| (b. rand, constant 1990 prices) | | | | | | |
| % change | | – 4.4 | – 4.7 | – 5.4 | 5.1 | 5.9 |
| as share of GDP (%) | 27.2 | 26.1 | 25.1 | 24.3 | 25.2 | 26.0 |
| Government deficit | 12.5 | 9.3 | 13.8 | 25.3 | 28.4 | 16.0 |
| (b. rand, constant 1990 prices) | | | | | | |
| as share of expenditure (%) | 14.1 | 11.3 | 16.6 | 28.0 | 29.2 | 18.0 |
| as share of GDP (%) | 4.5 | 3.4 | 5.0 | 9.5 | 10.5 | 5.7 |
| Total government debt | 91.9 | 96.0 | 93.0 | 99.2 | 109.2 | 123.8 |
| (b. rand, constant 1990 prices) | | | | | | |
| as share of GDP (%) | 33.2 | 34.8 | 34.1 | 37.1 | 40.3 | 44.5 |

*Sources:* South African Department of Finance, *Budget Review 1995* (South African Department of Finance: Pretoria, 1995); and South African Reserve Bank, *Quarterly Bulletin* (South African Reserve Bank: Pretoria, various issues).

The declining performance of the South African economy after 1989 was primarily a function of the deteriorating performance of the agricultural, mining and manufacturing sectors. The poor performance of the manufacturing sector was related to the declining levels of investment and factor productivity.[70] However, despite the economy's poor performance in 1989–93, South Africa's external position improved quite significantly after 1989 as a result of the gradual lifting of trade and financial sanctions which accompanied the country's progress towards democracy.

Government economic policy and spending priorities changed significantly after 1989 in the context of the changing political situation and the recession. The appointment of Derek Keys (formerly of Gencor) as Finance Minister in 1992 coincided with the implementation of the government's new economic strategy, the Normative Economic Model, which combined supply- and demand-side measures in order to increase growth and employment, control government expenditure, reduce inflation and redistribute resources to the poorest sections of the population.[71] These macroeconomic reforms included major reductions in defence expenditure as part of the attempt to hold public expenditure down. The government's fiscal policies also exhibited marked shifts between 1989 and 1993, with the major emphasis being on reducing the budget deficit and constraining government current expenditure (see table 3.2).

[70] Joffe *et al.* (note 69), pp. 3–12.
[71] *South Africa Country Report 1992/93* (Economist Intelligence Unit: London, 1993), p. 19.

Despite tighter fiscal and monetary policies, the government was unable to contain the increases in public expenditure, particularly current expenditure, between 1989 and 1993. Although government expenditure fell in 1990, it increased quite significantly in 1992 and 1993. Revenue was declining as a result of the recession, and this was reflected in the increasing size of the deficit and total government debt (see table 3.2). Total government debt as a percentage of GDP increased from 33 per cent in 1989 to 44.5 per cent in 1994, although this was still better than those of many Organisation for Economic Co-operation and Development (OECD) countries. However, the government deficit as a percentage of GDP increased from 4.5 per cent in 1989 to over 10 per cent in 1993, which was high in comparison with most OECD countries.[72]

Two disturbing trends in the structure of government expenditure were the falling share of capital expenditure[73] as current expenditure continued to absorb an increasing amount of the national budget and the rising share of interest payments on state debt, which accounted for 15 per cent of total government expenditure in 1993, up from 13 per cent in 1989.[74]

It is notable that cuts in military expenditure, often perceived as a disarmament measure, were brought about by macroeconomic pressures rather than by any political desire to curtail the flow of resources to the military.

Another feature of government economic policy which had a significant impact on the economic environment after 1989 and was to influence the future shape and structure of the South African defence industrial base was the issue of privatization and deregulation. In the post-World War II period up until the 1980s the state had become involved in a number of 'strategic' industries, such as synthetic fuels (Sasol), armaments (Armscor) and oil exploration (Mossgas). However, after de Klerk came to power in 1989, privatization and deregulation were seen as essential elements of the government's policy to 'free up' the economy and promote economic growth. In late 1989 Iscor, the government-owned iron and steel producer, was privatized and listed on the Johannesburg Stock Exchange. In addition a number of state corporations—Telkom, Transnet, Eskom, Sasol and Armscor (Denel)—were 'commercialized' during the early 1990s in order to make them more commercially viable and profitable in preparation for possible privatization.

The issue of privatization became intensely controversial in the early 1990s as many opposition parties (such as the ANC) felt that the government was unilaterally restructuring or selling off large parts of the public sector before the advent of majority rule. As a result the government backed down on the issue of privatization of state corporations and placed the emphasis on commercializa-

---

[72] The average government deficit as a percentage of GDP among the OECD countries in 1993 was less than 3% except in Sweden (8.5%), the UK (3.4%) and Australia (3.35%). *Financial Mail*, 20 May 1994.

[73] Capital expenditure as a share of total government expenditure declined from 13.6% in 1989 to 5.9% in 1993.

[74] This is significantly above the 7.1% average for OECD countries in 1993. *Financial Mail*, 20 May 1994.

tion and deregulation in an attempt to make the public sector more efficient.[75] The government also attempted to reduce the power of cartels and monopoly producers in the economy by giving the Competition Board new powers to block mergers and acquisitions which were detrimental to the public interest in that they enhanced the position of existing monopolies and conglomerates.

## V. Conclusions

The changes in South Africa's external strategic environment were dominated by the end of the cold war and of superpower rivalry in many parts of the world, including Southern Africa. The withdrawal of its armed forces from Angola and Namibia, the resolution of many of its historical conflicts and the move towards political pluralism in the region also had a significant impact on its strategic environment. The changes in the domestic political setting, particularly the abolition of the last vestiges of institutionalized apartheid, were largely the result of de Klerk's decision to implement a series of fundamental political reforms and of the ANC's acceptance of the negotiation option for finding a solution to the country's political crisis. The deterioration in South Africa's economic situation between 1989 and 1994, as a result of the severe domestic recession and the economic legacies of apartheid, had a number of contradictory effects. On the one hand it hastened the process of political reform; on the other hand it undermined the positive developments taking place on the political front by exacerbating the country's many socio-economic problems, such as high levels of unemployment and the great disparities between blacks and whites with respect to wealth, income and access to basic resources.

The period 1989–94 marked a major transition in South Africa's political economy, with a liberalization of the body politic and the implementation of a process of disarmament. The partial demilitarization of the state and the macroeconomic adjustment policies implemented by de Klerk were not a conscious attempt to reduce the power of the military establishment. However, they inadvertently initiated the process of deconstructing the military–industrial complex that had been built up during the apartheid era.

[75] *South Africa Country Report 1992/93* (note 71), p. 19.

# 4. Defence cuts and disarmament measures, 1989–94

## I. Introduction

South Africa's transition to democracy between 1989 and April 1994 was accompanied by a process of disarmament which included dramatic cuts in defence expenditure, reductions in force levels and the implementation of various disarmament measures. Budgetary constraints (the result of the severe domestic recession) and the dramatic changes in South Africa's external strategic environment described in chapter 3 were the two most significant determinants of the defence cuts and disarmament measures.

During the 1989–94 period of disarmament, President F. W. de Klerk initiated a number of 'political reforms' which included the unbanning of opposition parties and the release of political prisoners. While these reforms contributed to the liberalization of state security policies and the establishment of civilian control over the armed forces, they did not challenge many of the most significant institutional and industrial manifestations of militarism from the apartheid era, such as the power and influence of the local military–industrial complex. Nevertheless, the defence cuts did have a significant impact on the size and structure of the domestic arms industry, as reflected in declining sales, output, employment and profitability. This chapter describes the nature and extent of those defence cuts and disarmament measures and their impact on the domestic arms industry and the national economy.

## II. Defence cuts

In contrast to the militarization period 1975–89, when it showed sustained real increases, after 1989 the defence budget was cut quite dramatically—by over 40 per cent in real terms between 1989 and 1993. The reduction is reflected in the decline in South Africa's 'military burden', as measured by the share of military expenditure in GDP and total government expenditure (see table 4.1).

The cuts in defence expenditure were accompanied by significant changes in the structure of the defence budget (see table 4.2). The four main programmes, landward defence (army), air defence (air force), maritime defence (navy) and medical support, increased their shares of the total budget. The shares of programmes such as the General Support Programme fell as a result of the rationalization of various functions as a result of the end of South Africa's involvement in the war in Angola and Namibia.[1]

---

[1] The Quartermaster General (QMG) organization was disbanded in 1990 and the purchasing functions of the QMG transferred to the individual services.

**Table 4.1.** Military expenditure, 1989–94

|  | 1989 | 1990 | 1991 | 1992 | 1993 | 1994 |
|---|---|---|---|---|---|---|
| Military expenditure (m. rand, constant 1990 prices) | 11 435 | 10 070 | 8 094 | 7 605 | 6 589 | 7 153 |
| % change on previous year | *5.7* | *– 11.9* | *– 19.6* | *– 6.0* | *– 13.4* | *8.6* |
| Total government exp. (m. rand, constant 1990 prices) | 87 913 | 81 380 | 82 480 | 90 343 | 96 718 | 88 374 |
| GDP (m. rand, constant 1990 prices) | 276 940 | 276 060 | 273 249 | 267 257 | 270 702 | 278 148 |
| Mil. exp. as share of GDP (%) | *4.1* | *3.6* | *3.0* | *2.8* | *2.4* | *2.6* |
| Military exp. as share of government exp. (%) | *13.0* | *12.4* | *9.8* | *8.4* | *6.8* | *8.1* |

*Sources:* South African Department of State Expenditure, *Printed Estimates of Expenditure* (South African Department of State Expenditure: Pretoria, various years); and South African Department of Finance, *Budget Review* (South African Department of Finance: Pretoria, various years).

The structure of the defence budget by function also showed significant changes between 1989 and 1993 (see table 4.3). The share of personnel and operating costs increased at the expense of procurement and R&D. The increasing share of personnel expenditure was related to: (*a*) improved conditions of service and pension contributions for SADF personnel as well as the costs associated with laying off of staff, and (*b*) the internal deployment of the SADF in support of the police, particularly as a result of the increasing political violence which accompanied the final stages of the process of constitutional negotiations before the elections in April 1994.[2] The dramatic declines in the share of procurement and R&D expenditure (see table 4.3) and in the Special Defence Account (see table 4.2) were related to the cancellation and postponement of armaments projects[3] and the curtailment of 'special defence activities', including military intelligence operations.[4]

Despite these significant cuts, the country remained the largest military spender on the continent in absolute terms and accounted for nearly 65 per cent of total military expenditure in Southern Africa and 27 per cent of total military expenditure in Africa.[5] However, South Africa's military burden, as measured by military expenditure as a proportion of GDP, was lower than those of many other countries in Southern Africa, including Angola (32.8 per cent), Botswana (4.1 per cent), Mozambique (7.6 per cent) and Zimbabwe (3.8 per cent).[6]

---

[2] For details of the process of constitutional negotiations, see Friedman, S. and Atkinson, D., *South African Review 7: The Small Miracle. South Africa's Negotiated Settlement* (Ravan Press: Johannesburg, 1994).

[3] On the Special Defence Account, see chapter 2, section V in this volume.

[4] Private communication with personnel from the Chief of Staff (Finance) of the SADF, 7 Apr. 1994.

[5] George, P. *et al.*, 'Military expenditure', *SIPRI Yearbook 1996: Armaments, Disarmament and International Security* (Oxford University Press: Oxford, 1996), pp. 359–78.

[6] George *et al.* (note 5), pp. 373–75.

**Table 4.2.** The distribution of military expenditure by main programmes, 1989–94
Figures are percentages.

|  | 1989 | 1990 | 1991 | 1992 | 1993 | 1994 |
|---|---|---|---|---|---|---|
| Command and control | 2.0 | 2.1 | 2.8 | 2.9 | 3.1 | 3.7 |
| Landward defence | 12.6 | 17.6 | 22.0 | 22.6 | 26.0 | 37.0 |
| Air defence | 10.3 | 11.8 | 15.8 | 15.5 | 15.4 | 14.9 |
| Maritime defence | 3.8 | 4.2 | 5.0 | 5.4 | 6.4 | 5.9 |
| Medical support | 2.8 | 3.2 | 4.2 | 4.3 | 5.4 | 7.7 |
| General support | 10.0 | 4.0 | 4.8 | 4.1 | 3.6 | 3.0 |
| Special Defence Account[a] | 58.5 | 57.1 | 45.4 | 45.2 | 40.1 | 27.9 |
| **Total** | **100.0** | **100.0** | **100.0** | **100.0** | **100.0** | **100.0** |

[a] All procurement expenditure is funded from the Special Defence Account.

*Source:* South African Department of State Expenditure, *Printed Estimates of Expenditure* (South African Department of State Expenditure: Pretoria, various years).

## III. Disarmament measures

The cuts in defence expenditure which were implemented between 1989 and 1993 were achieved as a result of, and in conjunction with, a wide variety of disarmament measures. These measures, which are examined in detail below, were implemented not only in order to realize savings on defence but also in response to the changing political and strategic environment. They also made a significant contribution to the process of disarmament which accompanied South Africa's transition to democracy.

### Rationalization and restructuring of the SADF

The entire organizational structure and infrastructure of the SADF were rationalized and reorganized between 1989 and 1993. This included the ending of conscription for white males, the disbanding of various SADF units, and the closure, scaling-down or rationalization of various military bases and installations (see appendix 1). In December 1989 President de Klerk announced that national service for white males was being cut from two years to one year from the beginning of 1990.[7] In August 1993 the Minister of Defence, Kobie Coetzee, announced that the system of compulsory military service for white men would be abolished at the end of 1993 and replaced by a voluntary system.[8] The bulk of the rationalization measures affected the navy and air force, and were concentrated in the Western Cape. The air force closed five bases and disbanded 10 squadrons between 1989 and 1993. The army and medical service

---

[7] *Race Relations Survey 1989/90* (South African Institute for Race Relations: Johannesburg, 1990), p. 137.
[8] *The Citizen*, 25 Aug. 1993.

**Table 4.3.** The distribution of military expenditure by function, 1989–94

Figures are percentages.

|  | 1989 | 1990 | 1991 | 1992 | 1993 | 1994 |
|---|---|---|---|---|---|---|
| Personnel | 18.9 | 21.2 | 27.7 | 27.9 | 31.6 | 38.6 |
| Operating | 22.6 | 21.7 | 26.9 | 26.9 | 28.3 | 33.5 |
| Procurement[a] | 58.5 | 57.1 | 45.4 | 45.2 | 40.1 | 27.9 |
| Total | 100.0 | 100.0 | 100.0 | 100.0 | 100.0 | 100.0 |
| Research and development[b] | 8.6 | 7.9 | 7.2 | 6.1 | 5.2 | 4.8 |

[a] All procurement expenditure is funded from the Special Defence Account.

[b] R&D expenditure is funded from operating (General Support Programme) and procurement (Special Defence Account) expenditure.

*Sources:* South African Department of State Expenditure, *Printed Estimates of Expenditure* (South African Department of State Expenditure: Pretoria, various years); and private communications with Armscor officials.

were also rationalized, but to a lesser degree. Most of the army and medical service rationalization measures affected regions such as the Eastern Cape (Port Elizabeth), the Pretoria–Witwatersrand–Vereeniging (PWV) region (Pretoria) and the North-West (Potchefstroom).

## Cancellation or postponement of weapon projects

In 1990 it was announced that 11 major weapon and equipment projects had been cancelled while a further 49 had been postponed, delayed or spread out over longer time-periods.[9] In the following years few new major procurement projects were initiated (the purchase of new trainer aircraft from Switzerland was an exception). The share of the Special Defence Account programme fell from nearly 60 per cent in 1989 to 40 per cent in 1993.

## Cuts of SADF and Armscor personnel

Rationalization and restructuring of the SADF together with the ending of conscription resulted in large-scale reductions in SADF personnel after 1989. Between 1989 and 1993 over 37 000 SADF full-time personnel were cut, including 28 000 conscripts and 9000 permanent force members.[10] The total size of the SADF's active full-time personnel, including civilians, declined from *c.* 115 000 in 1989 to *c.* 78 000 in 1993.[11] The cuts in the defence budget, particularly in procurement, had a negative impact on Armscor's procurement

[9] *Financial Mail,* 21 Sep. 1990, p. 24.

[10] *The Star* (Johannesburg), 28 June 1994; and *Business Day,* 24 Aug. 1994.

[11] These figures are based on private communications with personnel from Chief of Staff (Personnel), SADF, Pretoria, Apr. 1994.

and production activities, and almost 10 000 personnel were laid off by Armscor and Denel between 1989 and 1993.

## Withdrawal and/or sale of redundant, obsolete and surplus equipment

The rationalization and restructuring of the SADF after 1990 meant that several types of equipment were withdrawn from service. This was particularly the case with the air force, where certain aircraft (such as the Buccaneer) were withdrawn from service after 1990.[12] Nearly 400 aircraft were withdrawn from service between 1989 and 1993: the total number of aircraft was reduced from 780 to c. 380.[13] The SADF sold large amounts of surplus military equipment after 1989 in an attempt to offset the severe defence cuts. Armscor, which acts as an agent for the SADF, sold redundant and surplus equipment worth 46.5 million rand in 1992 and 74 million rand in 1993.[14] Sales of surplus SADF equipment accounted for nearly 10 per cent of South Africa's total arms exports in 1992 and 1993.[15]

## Termination of the nuclear weapon programme

In the improved strategic and political environment, South Africa's nuclear weapon programme, started in the early 1970s, was terminated by de Klerk at the end of 1989.[16] In March 1993, nearly two years after accession to the NPT, de Klerk admitted to parliament that South Africa had constructed six complete nuclear devices and one incomplete device between 1980 and 1989. He also stated that the nuclear weapon programme had been abandoned in view of the normalization of South Africa's international relations and a changed global political situation in which 'a nuclear deterrent had become not only superfluous, but in fact an obstacle to the development of South Africa's international relations'.[17] According to information obtained from Armscor, the pilot enrichment plant at Pelindaba was closed in early 1990, and this was followed by the dismantling of the seven gun-type nuclear devices. Thereafter the highly enriched uranium (HEU) used in the devices was melted down and recast into a form that made it unsuitable for use in a nuclear weapon. It was then returned to the Atomic Energy Corporation at Pelindaba where it was stored in accordance with International Atomic Energy Agency (IAEA) guidelines. The weapon manufacturing facility at Advena, which became part of Denel in 1992, was decontaminated, and the test-shafts at Vastrap in the Northern Cape were back-

[12] For a detailed list of the equipment withdrawn from service in the Air Force, see *South African Defence Force Review* (Department of Defence: Pretoria, 1990), pp. 93–107.

[13] *The Star* (Johannesburg), 2 June 1993.

[14] *Armscor Annual Report 1992/93*, p. 16; and *1993/94*, p. 20.

[15] *Armscor Annual Report 1992/93, 1993/94*.

[16] For a detailed discussion of South Africa's nuclear weapon programme, see Albright, D., 'The legacy of the South African nuclear weapons program', Paper presented at the Conference on a Nuclear Policy for a Democratic South Africa, Cape Town, 11–13 Feb. 1994.

[17] South African Parliament, *Hansard*, 24 Mar. 1993, cols 3465–71.

**Table 4.4.** Domestic arms production, 1989–94

|  | 1989 | 1990 | 1991 | 1992 | 1993 | 1994 |
|---|---|---|---|---|---|---|
| Military expenditure (m. rand, constant 1990 prices) | 11 435 | 10 070 | 8 094 | 7 605 | 6 589 | 7 153 |
| Domestic arms production[a] (m. rand, constant 1990 prices) | 4 117 | 3 685 | 4 093 | 3 079 | 3 250 | 2 643 |
| Value of arms production (1990=100)[b] | 111.7 | 100.0 | 111.1 | 83.6 | 88.2 | 71.7 |
| Volume of arms production (1990=100)[c] | 111.5 | 100.0 | 90.5 | 70.6 | 63.5 | 63.2 |
| Arms production as share of manufacturing output (%) | 6.6 | 5.8 | 6.7 | 5.3 | 5.7 | 4.5 |
| Arms prod. as share of GDP (%) | 1.5 | 1.3 | 1.5 | 1.3 | 1.2 | 1.0 |

[a] Production = total procurement for the SADF, the police and other government departments, minus imports plus exports.

[b] 1990 = 100.

[c] Estimated from total arms industry employment (assumes no changes in productivity). 1990 = 100.

*Sources:* South African Department of State Expenditure, *Printed Estimates of Expenditure* (South African Department of State Expenditure: Pretoria, various years); private communications with Armscor officials; and South African Reserve Bank, *Quarterly Bulletin* (South African Reserve Bank: Pretoria, various issues).

filled with sand and concrete in mid-1993.[18] The IAEA visited South Africa's nuclear facilities in late 1993 and reported on its nuclear weapon programme.[19]

The cost of the nuclear weapon programme between the early 1970s and 1989, according to de Klerk, was estimated at 700–800 million rand in current prices. This has been disputed by certain observers, who put it at *c.* 8 billion rand.[20]

As a result of South Africa's accession to the NPT and to the Chemical Weapons Convention (CWC) in early 1993 and of the termination of the nuclear weapon programme, new 'non-proliferation' legislation was passed in 1993. This included the Non-Proliferation of Weapons of Mass Destruction Act no. 87 of 1993 and the Space Affairs Act no. 84 of 1993.

## IV. The economic impact

The defence cuts and disarmament measures implemented after 1989, coinciding with the deteriorating performance of the manufacturing sector and the national economy (see table 3.1), had a number of short-term economic costs, such as large-scale cuts in numbers of defence workers and the down-sizing and restructuring of the domestic arms industry.

[18] Private communication with Armscor official, Apr. 1994.
[19] Howlett, D. and Simpson, J., 'Nuclearisation and denuclearisation in South Africa', *Survival*, vol. 35, no. 3 (1993), pp. 154–73.
[20] *Weekly Mail*, 26 Mar. 1993, p. 5.

## Arms production

Procurement expenditure by the SADF fell by over 60 per cent in real terms between 1989 and 1993. At an aggregate level the value of arms production declined by 20 per cent and the volume of arms production by 43 per cent over the same period. What is evident, however, is that neither of these declines was as dramatic as the fall in SADF procurement expenditure. Declining demand for armaments from the SADF after 1989 was offset to some extent by increased exports and increased demand from the police and other government departments such as the correctional services, but even so the contribution of the arms industry to total manufacturing output declined from 6.6 per cent in 1989 to 4.5 per cent in 1994, while the share of domestic arms production in GDP declined from 1.5 per cent in 1989 to 0.95 per cent in 1994.

## Military research and development

The cuts in defence expenditure were accompanied by significant cuts in military R&D. At an aggregate level, this declined by 65 per cent in real terms between 1989 and 1993. The fall was reflected in the decline in the share of military R&D in total R&D spending—from 48 per cent in the late 1980s to about 18 per cent in 1993. In the context of the dramatic cuts in procurement, a larger share of total military R&D expenditure was spent on technology development projects as opposed to the R&D components of procurement projects (see table 4.5). This was the result of a desire to retain key defence technologies and capabilities within the local arms industry.

The impact of the decline in military R&D expenditure was felt most acutely in terms of human capital resources. Armscor shed almost 40 per cent of its engineers and scientists (nearly 800) between 1989 and 1992, and these trends were reflected in the private sector, according to interviews with private-sector company officials.[21]

While there is no publicly available information as to where these human resources ended up after they were released from the defence sector, certain trends can be identified from information given in interviews with personnel from the public- and private-sector arms industry. Many of the scientists and engineers working in the military R&D sector emigrated or moved into non-military R&D markets. The emigration of scientists and engineers is evident from the declining numbers of engineers and scientists as a share of the total labour force.[22]

Although these figures do not provide clear trends, it is possible that some of the resources previously allocated to military R&D were reallocated to other forms of civilian R&D (both private and public). However, despite the real

[21] Private communication with A. Holloway, Grinaker Avitronics, May 1996.

[22] Kaplan, D., 'Ensuring technological advance in the South African manufacturing industry: some policy proposals', in A. Joffe, D. Kaplan, R. Kaplinsky and D. Lewis, *Improving Manufacturing Performance in South Africa* (University of Cape Town Press: Cape Town, 1995), p. 239.

**Table 4.5.** Military research and development expenditure, 1989–94

|  | 1989 | 1990 | 1991 | 1992 | 1993 | 1994 |
|---|---|---|---|---|---|---|
| Military R&D[a] | 985 | 793 | 580 | 467 | 343 | 342 |
| (m. rand, constant 1990 prices) |  |  |  |  |  |  |
| Technology development[b] | 180 | 163 | 340 | 321 | 210 | 188 |
| (m. rand, constant 1990 prices) |  |  |  |  |  |  |
| as % of total | *18.2* | *20.6* | *58.7* | *68.8* | *61.2* | *55.0* |
| Procurement projects | 805 | 630 | 240 | 146 | 132 | 154 |
| (m. rand, constant 1990 prices) |  |  |  |  |  |  |
| as % of total | *81.8* | *79.4* | *41.3* | *31.2* | *38.8* | *45.0* |
| Military expenditure | 11 435 | 10 070 | 8 094 | 7 605 | 6 589 | 7 153 |
| (m. rand, constant 1990 prices) |  |  |  |  |  |  |
| R&D as share of mil. exp. (%) | *8.6* | *7.9* | *7.2* | *6.1* | *5.2* | *4.8* |
| Total R&D | 2 043 | .. | 2 455 | .. | 1 831 | .. |
| (m. rand, constant 1990 prices) |  |  |  |  |  |  |
| Military R&D as share of total R&D (%) | *48.2* | .. | *23.6* | .. | *18.7* | .. |
| Total R&D as share of GDP (%) | *0.7* | .. | *0.9* | .. | *0.7* | .. |

[a] Excludes military R&D funded by private-sector defence companies. R&D expenditure is funded from operating (General Support Programme) and procurement (Special Defence Account) expenditure.

[b] Includes the Industry and Technology Survival Plan (ITSP) and the Technology Development Programme (TDP).

*Sources:* South African Department of State Expenditure, *Printed Estimates of Expenditure* (South African Department of State Expenditure: Pretoria, various years); private communications with Armscor officials; and *SA Science and Technology Indicators 1996* (Foundation for Research Development: Pretoria, 1996).

increases in the level of total R&D expenditure between 1989 and 1991, as a percentage of GDP total R&D expenditure in 1993 (0.68 per cent) was less than in the late 1980s (0.74 per cent in 1989). This was in sharp contrast with the trend in many successful developing economies which witnessed real and sustained increases in R&D expenditure during the late 1980s and early 1990s.[23]

It is extremely difficult to ascertain whether the reallocation of resources from defence to civilian R&D had a positive impact on the performance, productivity and competitiveness of the civilian economy between 1989 and 1993. Certain indicators showed positive signs after 1989. South Africa's share of the world publication output increased from 0.52 per cent in 1981–85 to 0.56 per cent in 1990–94, and the number of patents granted in South Africa increased from 5597 in 1989 to 6129 in 1992. More importantly, the number of South African patents granted in South Africa increased from 1211 in 1989 to 1417 in 1992, providing some evidence of increased R&D activity and scientific output.

[23] South Africa's share of R&D spending in GDP (0.67%) in 1993 was significantly less than that of certain developing countries such as South Korea (2.08%), Singapore (1.18%) and Taiwan (1.76%), but higher than those of others such as Malaysia (0.17%) and Mexico (0.37%). *SA Science and Technology Indicators 1993* (Foundation for Research and Development: Pretoria, 1993).

**Table 4.6.** Arms industry employment, 1989–94

|                                              | 1989    | 1990    | 1991    | 1992    | 1993    | 1994    |
| -------------------------------------------- | ------- | ------- | ------- | ------- | ------- | ------- |
| Total employment (th.)[a]                    | 5 650.2 | 5 633.3 | 5 537.5 | 5 424.9 | 5 312.2 | 5 278.4 |
| % change                                     |         | – 0.3   | – 1.7   | – 2.0   | – 2.1   | – 0.6   |
| Manufacturing employment (th.)               | 1 583.3 | 1 581.7 | 1 546.9 | 1 504.2 | 1 477.3 | 1 480.5 |
| % change                                     |         | – 0.1   | – 2.2   | – 2.8   | – 1.8   | 0.2     |
| Total arms ind. employment (th.)[b]          | 131.8   | 118.2   | 106.9   | 83.4    | 75.0    | 74.6    |
| % change                                     |         | – 10.3  | – 9.5   | – 22.0  | – 10.1  | – 0.5   |
| Armscor/Denel (no. of staff, th.)            | 26.4    | 23.6    | 21.4    | 16.4    | 15.0    | 14.9    |
| Arms industry as share of total employment (%) | 2.3   | 2.1     | 1.9     | 1.5     | 1.4     | 1.4     |
| Arms industry as share of manufacturing employment | 8.3 | 7.5   | 6.9     | 5.5     | 5.1     | 5.0     |
| Government employment (th.)                   | 1 707.3 | 1 724.6 | 1 766.0 | 1 786.7 | 1 764.3 | 1 774.6 |

[a] Total non-agricultural employment.

[b] Includes all staff in private- and public-sector defence companies involved in both defence and non-defence work.

*Sources:* Private communication with Armscor officials, Pretoria, Apr. 1994; and South African Reserve Bank, *Quarterly Bulletin* (South African Reserve Bank: Pretoria, various issues).

However, despite these improvements in scientific output after 1989, the performance and competitiveness of the South African economy, and particularly the manufacturing sector, deteriorated quite significantly between 1989 and 1993. The impact of the defence cuts on the manufacturing sector and the reasons for the poor performance of the manufacturing sector between 1989 and 1993 are examined below.

There were no major changes in the way in which military R&D was funded in the period 1989–93. It continued to be funded out of the General Support Programme and the Special Defence Account of the defence budget. The DRDC's role remained the same. Armscor, as the SADF's procurement agency, continued to place military R&D contracts in both the public- and the private-sector firms. Most private-sector defence companies also funded certain R&D activities from their own resources. With the establishment of Denel in April 1992 a number of R&D activities which had previously been carried out in Armscor subsidiaries were transferred to Denel's 'new' divisions and subsidiary companies.

## Employment

The defence cuts and disarmament measures after 1989 had a significant impact on employment in the defence sector, particularly in the arms industry. Total employment in the domestic arms industry declined by nearly 60 000 between 1989 and 1993, while the total number of full-time members of the SADF fell

**Table 4.7.** Armscor and Denel: distribution of job losses, 1989–92

|                      | Total 1989 | % of total | Total 1992 | % of total | % change 1989–92 |
|----------------------|------------|------------|------------|------------|------------------|
| Engineers/scientists | 1 987      | 7.5        | 1 181      | 7.2        | − 40.6           |
| Computer specialists | 1 160      | 4.4        | 835        | 5.1        | − 28.0           |
| Technicians          | 3 164      | 12.1       | 2 196      | 13.3       | − 30.6           |
| Admin./finance       | 6 125      | 23.2       | 4 434      | 27.1       | − 27.6           |
| Artisans             | 3 671      | 13.9       | 2 276      | 13.8       | − 38.0           |
| Labourers/other      | 10 241     | 38.9       | 5 490      | 33.5       | − 46.4           |
| **Total**            | **26 348** | *100.0*    | **16 411** | *100.0*    | − 37.7           |

*Source:* Private communication with Armscor and Denel officials, Apr. 1994.

by 37 000.[24] In all almost 100 000 jobs were lost in the defence sector over this period. The job losses in the arms industry corresponded with job losses in the manufacturing sector—it shed over 100 000 jobs (a 6 per cent decline) between 1989 and 1993. Total non-agricultural employment fell by nearly 340 000 over the same period. The government sector was the only sector of the economy which experienced a slight increase in total employment after 1989. Employment in the public-sector arms industry declined by over 43 per cent between 1989 and 1993 (see table 4.6).

It is evident that the job losses in the defence sector after 1989 were not offset by compensating increases in employment in other, civil sectors of the economy. Thus the cuts in defence expenditure may have contributed to a short-run net increase in unemployment, both in the manufacturing sector and in the economy as a whole. The lack of job creation in the civil sector was linked to a number of factors including the absence of government macroeconomic compensatory policies and the recession. There is no evidence to suggest that the government used the savings from the defence cuts to fund retraining schemes for defence workers or to create new civilian markets which could have provided work for former defence workers. Company officials also highlighted the fact that the high degree of secrecy which characterized the domestic arms industry during the apartheid era, and which inhibited the creation of linkages between the arms industry and other sectors of the manufacturing base, also limited the ability of defence workers to find alternative employment opportunities in the civil sector.

According to information obtained from Armscor, the distribution of job losses within the public-sector arms industry tended to favour computer specialists and administrative personnel at the expense of engineers, scientists, artisans and labourers (see table 4.7). Disaggregated information on the distribution of job losses in the private-sector arms industry is not readily available,

---

[24] *Race Relations Survey 1994/95* (South African Institute for Race Relations: Johannesburg, 1995), pp. 153–54; and see note 10.

**Table 4.8.** Armscor: regional distribution of job losses, 1989–92

| Location | No. of jobs lost | % of total jobs lost |
|---|---|---|
| Western Cape | 1 563 | 21.7 |
| Orange Free State | 17 | 0.2 |
| PWV | 5 613 | 78.1 |
| **Total** | **7 193** | *100.0* |

*Source:* Private communication with C. Hoffman, General Manager, Finance and Administration, Armscor, Apr. 1994.

but information obtained from interviews with company officials suggests that it followed similar patterns to that in the public-sector arms industry, with engineers, scientists and unskilled labourers bearing the brunt of the job losses. Company officials also suggested that significant numbers of skilled workers (mainly engineers and scientists) from the private-sector arms industry had emigrated since 1989, further depleting the country's scarce supply of skilled labour. The loss of skilled workers, even if only a small proportion of the total supply of skilled workers, constituted an unforeseen cost of disarmament.

**Location effects**

The negative short-run effects of defence cuts and disarmament measures, particularly in terms of the underutilization of resources, tend to be concentrated in those towns and/or regions which are heavily dependent upon defence expenditure as a result of the location of military bases and/or production facilities. In South Africa the location effects were directly related to the geographical distribution of military bases and defence production facilities (see table 4.8).

The bulk of the job losses in Armscor between 1989 and 1992 were concentrated in the PWV region, over 21 per cent in the Western Cape and less than 1 per cent in the Orange Free State (OFS). These cuts corresponded with the geographical location of Armscor production facilities—60 per cent in the PWV region, 33 per cent in the Western Cape and 6 per cent in the OFS. Detailed information on the geographical distribution of job losses in the private-sector arms industry is not available. However, the geographical distribution of the private-sector arms industry corresponds with the distribution of Armscor production facilities, and it is therefore possible to assume that the location effects of the defence cuts on the private-sector arms industry mirrored those in the public-sector arms industry.[25]

[25] For a discussion of the geographical distribution of arms production in South Africa, see Rogerson, C., 'Defending apartheid: Armscor and the geography of military production in South Africa', *GeoJournal*, vol. 22, no. 3 (1990), pp. 241–50. See also *The Armaments Industry in South Africa* (Armscor: Pretoria, 1984) for details of the geographical distribution of Armscor contractors; and chapter 5, section IV in this volume.

**Table 4.9.** Regional economic indicators, 1993

Figures are regions' percentage shares in national total.

|  | Western Cape | OFS | Natal | PWV |
|---|---|---|---|---|
| Population | 8.9 | 6.8 | 20.9 | 16.8 |
| Labour force | 11.5 | 7.9 | 19.1 | 25.0 |
| Manufacturing employment | 22.8 | 8.9 | 24.8 | 23.1 |
| GDP | 12.5 | 5.7 | 14.1 | 37.6 |
| Manufacturing output | 11.0 | 2.8 | 16.3 | 45.0 |
| Number of manufacturing firms | 11.9 | 3.2 | 14.5 | 58.0 |

*Sources: South Africa's Nine Provinces: A Human Development Profile* (Development Bank of Southern Africa: Midrand, 1994); and South African Central Statistics Service, *South African Statistics Yearbook 1994* (Central Statistics Service: Pretoria, 1995).

The impact of the defence cuts and disarmament measures was particularly acute in the defence-dependent communities of the Western Cape region (e.g., Simonstown), given the scaling down and disbanding of several navy and air force bases and installations around Cape Town and the laying off of more than 1500 Armscor personnel. The bulk of the army and medical service rationalization measures and some of the air force rationalization measures affected the PWV region, given the concentration of defence bases and facilities around Pretoria, which was also the site of defence headquarters. The PWV region is the dominant region in terms of its share of the total labour force and its contribution to GDP (see table 4.9). It also has the highest concentration (nearly 60 per cent) of manufacturing establishments and high-technology industries, particularly in the defence-related sectors of the manufacturing sector.[26] In the absence of more specific data it is possible to assume that the PWV region was better able to absorb the impact of the defence cuts and disarmament measures than the other regions where arms production facilities were located.

## Government expenditure

The cuts in defence expenditure after 1989 were a response to South Africa's changing strategic, political and economic environment. They were also part of a deliberate government policy which aimed to reduce government spending while at the same time shifting resources towards social services such as education and health.[27]

The share of defence in total government expenditure declined significantly after 1989, and this coincided with increases in the shares of other categories of expenditure, such as police, education and health. The share of social services

---

[26] The defence-related part of the manufacturing sector includes iron and steel, non-ferrous metals, metal products, machinery, electrical machinery, motor vehicles and transport equipment.

[27] South African Department of Finance, *Budget Review 1995* (South African Department of Finance: Pretoria, 1995), pp. 2.9–2.14.

**Table 4.10.** Government expenditure by selected functions, 1989–94
Figures are percentages of total government expenditure.

| Category of expenditure | 1989 | 1990 | 1991 | 1992 | 1993 | 1994 |
| --- | --- | --- | --- | --- | --- | --- |
| Defence | 13.0 | 12.4 | 9.8 | 8.4 | 6.8 | 8.1 |
| Police | 3.8 | 5.6 | 6.4 | 6.2 | 6.7 | 6.9 |
| Education | 17.7 | 20.9 | 20.6 | 20.7 | 21.1 | 21.2 |
| Health | 9.0 | 10.1 | 11.0 | 10.8 | 10.6 | 10.5 |

*Sources:* South African Department of State Expenditure, *Printed Estimates of Expenditure* (South African Department of State Expenditure: Pretoria, various years); and South African Department of Finance, *Budget Review* (South African Department of Finance: Pretoria, various years).

(such as education, health and welfare) increased from 39 per cent of the total budget in 1989 to 44 per cent in 1993, while the share of protection services (such as police, justice and the correctional services) declined from over 20 per cent of the total budget in 1989 to 17 per cent in 1993.

It seems evident that there was almost a direct reallocation of resources from defence to police between 1989 and 1993, for the level and share of the police budget increased to the same extent as the level and share of defence declined (see table 4.10). This reallocation was largely a response to the rising levels of crime and political violence which accompanied the transition to democracy.

Despite the cuts in defence expenditure and government expenditure generally, the budget deficit, in absolute terms and as a percentage of total expenditure and GDP, did not fall significantly during the period when the defence budget was being cut. Between 1989 and 1993 it increased by 130 per cent in real terms from 12.5 billion rand to 28.4 billion in 1990 constant prices. As a proportion of government expenditure it increased from 14.1 per cent in 1989 to 29.2 per cent in 1993, and as a proportion of GDP from 4.5 per cent in 1989 to 10.5 per cent in 1993. Total government debt (including foreign debt) declined slightly in 1991 but then witnessed sustained real increases in the following years and reached 44.5 per cent of GDP in 1994 (see table 3.2). There is no evidence that the savings (nearly 5 billion rand in 1990 prices) achieved from the defence cuts between 1989 and 1993 were directly allocated to reducing the size of South Africa's budget deficit or total government debt, but it is possible that these would have been greater without the cuts in defence expenditure.

**Balance of payments**

The cuts in military expenditure, particularly procurement, led to significant falls in the value of arms imports. Between 1989 and 1993 it declined by over 76 per cent in real terms, largely as a result of the 60 per cent decline in procurement expenditure during the same period. The share of arms imports in

**Table 4.11.** The South African arms trade, 1989–94

|  | 1989 | 1990 | 1991 | 1992 | 1993 | 1994 |
|---|---|---|---|---|---|---|
| Arms imports | 2 691 | 1 982 | 847 | 618 | 644 | 421 |
| (m. rand, constant 1990 prices) | | | | | | |
| Manufactured imports | 45 121 | 39 942 | 40 424 | 37 992 | 42 194 | 53 668 |
| (m. rand, constant 1990 prices) | | | | | | |
| Arms as share of manufact. imports (%) | *6.0* | *5.0* | *2.1* | *1.6* | *1.5* | *0.8* |
| Total imports | 59 416 | 53 984 | 54 645 | 46 533 | 49 769 | 64 454 |
| (m. rand, constant 1990 prices) | | | | | | |
| Arms as share of total imports (%) | *4.5* | *3.7* | *1.5* | *1.3* | *1.3* | *0.7* |
| Arms exports | 236 | 163 | 700 | 382 | 625 | 550 |
| (m. rand, constant 1990 prices) | | | | | | |
| Manufactured exports | 26 440 | 24 652 | 23 404 | 19 970 | 19 711 | 22 214 |
| (m. rand, constant 1990 prices) | | | | | | |
| Arms as share of manufact. exports (%) | *0.9* | *0.7* | *3.0* | *1.9* | *3.2* | *2.5* |
| Total exports | 76 314 | 69 487 | 65 717 | 53 319 | 55 903 | 58 004 |
| (m. rand, constant 1990 prices) | | | | | | |
| Arms as % of total exports | *0.3* | *0.2* | *1.1* | *0.7* | *1.1* | *0.9* |
| Arms trade balance[a] | − 2 455 | − 1 819 | − 147 | − 236 | − 19 | 129 |
| (m. rand, constant 1990 prices) | | | | | | |

[a] Exports minus imports.

*Sources: Armscor Annual Report*, various years; and Central Statistics Service, *Bulletin of Statistics* (Central Statistics Service: Pretoria, various issues).

manufactured imports fell from 6 per cent to 1.5 per cent and their share in total imports from 4.5 per cent to 1.3 per cent (see table 4.11).

The cuts in procurement and the declining domestic defence market prompted local defence firms to pursue export markets vigorously after 1989. The introduction of the General Export Incentive Scheme (GEIS) in 1990 played a significant role in stimulating arms exports. As a result of these developments, the value of South Africa's arms exports increased by more than 160 per cent in real terms between 1989 and 1993, accounting for an increasing share of manufactured exports and total exports (see table 4.11).

The impact on the balance of payments was positive. As a result of the decline in the value of arms imports and the increasing value of arms exports, South Africa moved from a trade deficit in armaments of 2.4 billion rand in 1989 to a deficit of less than 20 million rand in 1993 in 1990 constant prices. In 1994 it registered a trade surplus in armaments for the first time.

# V. Defence cuts, arms production and the manufacturing sector

If a country possesses an arms industry, the impact of defence cuts may be compounded, depending on the size and structure of the arms industry and its

**Table 4.12.** Trends in arms production and total manufacturing output, 1989–94
1990 = 100.

| | 1989 | 1990 | 1991 | 1992 | 1993 | 1994 |
|---|---|---|---|---|---|---|
| *Armaments* | | | | | | |
| Value of production[a] | 111.7 | 100.0 | 111.1 | 83.6 | 88.2 | 71.7 |
| Volume of production[b] | 111.5 | 100.0 | 90.5 | 70.6 | 63.5 | 63.2 |
| *Manufacturing sector* | | | | | | |
| Value of production | 100.5 | 100.0 | 96.5 | 92.6 | 91.9 | 94.0 |
| Volume of production | 100.4 | 100.0 | 96.4 | 93.5 | 93.3 | 95.8 |
| Manufacturing capacity utilization (%) | 84.1 | 81.9 | 81.0 | 78.5 | 77.9 | 80.0 |
| Arms production as % of manufacturing output | 6.6 | 5.8 | 6.7 | 5.3 | 5.7 | 4.5 |
| Arms production as % of GDP | 1.5 | 1.3 | 1.5 | 1.2 | 1.2 | 1.0 |

[a] Calculated from the sum of domestic procurement and expenditure on maintenance of military equipment.

[b] Estimated from total arms industry employment (assumes no changes in productivity).

*Sources:* Armscor; South African Reserve Bank, *Quarterly Bulletin* (South African Reserve Bank: Pretoria, various issues); and Central Statistics Service, *Bulletin of Statistics* (Central Statistics Service: Pretoria, various issues).

links with the rest of the manufacturing sector. The release of resources (land, capital and labour) from the arms industry as a result of defence cuts may have negative short-run impacts on the output and performance of a country's manufacturing sector, depending on the state of the economy (i.e., new and alternative sources of demand), the presence or absence of government compensatory demand policies, and the supply responses of the civilian manufacturing sector. The extent of these negative short-run costs will also be contingent upon whether, and for how long, the former defence resources remain underutilized and on the ability of the civilian manufacturing sector to respond to the new and alternative sources of demand that may be created as a result of the defence cuts.

Chapter 2 shows how South Africa's arms industry emerged as a significant part of the country's industrial base during the apartheid years as a result of its privileged access to scarce national resources. In response to the UN embargo and the drive for self-sufficiency in armaments production, a large number of private-sector firms (over 2000) in the manufacturing sector were drawn into arms production as contractors, subcontractors and suppliers to Armscor. As a result significant inter-industry linkages between the arms industry and other parts of the manufacturing sector were established. Given the significance of the arms industry to the manufacturing sector it is possible to assume that the decline in domestic demand for armaments after 1989 had a negative short-run

**Table 4.13.** Sales in the electronics sector, 1989–92

|  | 1989 | 1990 | 1991 | 1992 |
|---|---|---|---|---|
| Total sales (m. rand, constant 1990 prices) | 12 975 | 12 092 | 11 339 | 10 862 |
| % change on previous year | −6.8 | −6.8 | −6.2 | −4.2 |
| Military sales (m. rand, constant 1990 prices) | 899 | 745 | 593 | 476 |
| % change on previous year | −17.9 | −17.1 | −20.4 | −19.7 |
| Military sales as share of total electronics sales (%) | 6.9 | 6.1 | 5.2 | 4.3 |
| Electronics as share of manufact. ind. sales (%) | 5.3 | 5.2 | 4.6 | 4.4 |

*Sources: Prospects for the Electronics Industry* (Business Marketing Intelligence: Johannesburg, 1992); and South African Central Statistics Service, *South African Statistics Yearbook 1994* (Central Statistics Service: Pretoria, 1995).

impact on the manufacturing sector as a whole, and particularly the defence-related sectors and sub-sectors of the manufacturing sector (see table 4.12).

The declines in the value and volume of arms production between 1989 and 1993 corresponded with the declines in the value of manufacturing sales and the volume of production in the manufacturing sector during the same period, although the fall in manufacturing sales (8 per cent by value in real terms) between 1989 and 1993 was less severe than the fall in the value of armaments production (21 per cent) during the same period.

Declining domestic demand for armaments after 1989 was reflected in the declining utilization of production capacity in the manufacturing sector as a whole. This suggests that many of the resources released from the defence sector remained underutilized during this period.

There is very little publicly available information on how the defence cuts after 1989 affected the share of defence business in each of the defence-related sectors and sub-sectors of the manufacturing sector. However, a study of the electronics sector undertaken by Business Market Intelligence (BMI) in 1992 provides some indication of the impact of the defence cuts there, and particularly on the military electronics sub-sector (see table 4.13).

Total sales in the electronics sector declined by 16 per cent in real terms between 1989 and 1992, which was significantly more than the 10 per cent decline in total manufacturing sales, and the share of the electronics sector in total manufacturing sales also fell. However, military electronics sales fell by much more—47 per cent in real terms between 1989 and 1992—and their share the electronics sector declined from nearly 7 per cent in 1989 to around 4 per cent in 1992.

Thus the declining value and volume of arms production after 1989 corresponded with declines in the value and physical volume of production for the manufacturing sector as a whole and for most of the defence-related sectors of the manufacturing sector. However, the fact that the declines in general manufacturing sales and in physical volume of production were less severe than the decline in the value and volume of arms production suggests that the cuts in defence expenditure and the declining domestic demand for armaments may have been compensated for by new and alternative sources of demand for civilian goods and services.

While the poor performance of the manufacturing sector between 1989 and 1993 may have been influenced by the defence cuts, a number of other important factors contributed. Gross domestic fixed investment as a percentage of GDP fell quite dramatically between 1989 and 1993, and by 1993 the share of investment in GDP had declined to around 15 per cent.[28] During the early 1990s South Africa's ratio of investment to GDP was less than half that of successful developing economies and was insufficient to replace plant and equipment as it wore out.[29]

## VI. Conclusions

This chapter examines the nature and impact of the defence cuts and disarmament measures in South Africa between 1989 and April 1994. They were implemented in the context of a severe domestic recession and therefore probably contributed to the economy's deteriorating performance during the same period. The defence cuts also had a significant impact on the domestic arms industry, reflected in declining sales and output and large-scale job losses, although there is some evidence that new and alternative sources of demand compensated in part.

As a result of the defence cuts and disarmament measures, the process of militarization which had occurred between 1975 and 1989 began to be reversed. However, despite the down-sizing of the domestic arms industry, many of the institutional manifestations of militarism from the apartheid era, such as the centrality and influence of the military–industrial complex within the South African economy, remained intact and only began to be seriously challenged in the post-apartheid period after April 1994.

[28] In the early 1980s, the share of investment in South Africa's GDP was above 25%. Joffe *et al.* (note 22), p. 10.

[29] Joffe *et al.* (note 22), pp. 11–12.

# 5. Defence industrial adjustment, 1989–94

## I. Introduction

South Africa's arms industry by the end of the 1980s had developed into one of the country's most significant industrial sectors in terms of employment, contribution to national income and number of firms involved. The economic significance of the arms industry was also enhanced by its substantial political influence within the state through the operation of a local military–industrial complex.

The substantial cuts in defence expenditure which were implemented in South Africa between 1989 and 1994 were part of a larger process of disarmament which had a profound impact on the size, structure and performance of the country's arms industry. The severity of the cuts not only reduced the ability of the state to fund its policy of support for the domestic arms industry. It also contributed to the gradual breakdown of the cohesiveness of the military–industrial complex.

In response to the cuts, the arms industry was forced to down-size and restructure, both at the aggregate level and at the level of individual firms. This process of down-sizing and restructuring was reflected in large-scale job losses, declining output and profitability, take-overs and mergers and the demise of a number of defence firms. Defence firms were forced to pursue a variety of supply-side adjustment strategies in order to deal with the sharply reduced domestic demand for defence goods and services.

The South African arms industry was not monolithic or homogenous. Firms differed quite considerably in terms of size, products, location, sector, orientation and ownership, and the adjustment strategies they adopted reflected the heterogeneous nature of the defence industrial base. This chapter examines the adjustment experiences of South Africa's arms industry between late 1989 and April 1994.[1]

## II. Defence industrial adjustment in the public-sector arms industry

Armscor by the end of the 1980s had developed into one of the largest and most significant state-owned companies in the country, largely as a result of its privileged access to state resources. Most significantly, Armscor was responsible for both arms production (through its various subsidiary companies) and procurement, and thus occupied the dominant position in the country's military–

---

[1] The information presented in this chapter is based on extensive personal interviews conducted with employees of public- and private-sector defence firms in South Africa.

industrial complex. The down-sizing and restructuring of the public-sector arms industry included the establishment of a new state-owned arms production company, Denel, in 1992.

## Armscor's adjustment strategies

By 1989 Armscor had nearly 30 000 employees and assets of 3.7 billion rand, and was ranked 15th in the country in terms of total employment and 30th in terms of total assets.[2] The organizational structure of Armscor consisted of a head office in Pretoria, nine research and testing facilities and 12 subsidiary companies, with the research facilities and subsidiary companies concentrated around Johannesburg, Pretoria and Cape Town.

An internal study undertaken in late 1991 which assessed Armscor's future role came to the conclusion that it would have to participate in the commercial market if it were to survive the impact of the recent defence cuts and disarmament measures. However, Armscor was prohibited from using its production facilities for commercial purposes by the Armaments Development and Production Act no. 57 of 1968 (as amended).[3] The government therefore decided to form a new industrial company and the outcome was the formation of Denel in April 1992. Denel inherited all Armscor's subsidiaries and research facilities, with the exception of the Alkantpan Ballistic Test Range in the Northern Cape and the Institute for Maritime Technology (IMT) at Simonstown in the Western Cape.

With the restructuring of the public-sector arms industry and the formation of Denel, Armscor still retained responsibility for procurement for the SADF. It also retained other functions such as marketing support, arms control (the issuing of export permits), the sale of SADF surplus weapons, and, most importantly, overall coordination of the local arms industry. After 1989, and particularly after 1992 in its 'new' role, Armscor pursued a number of adjustment strategies: (a) restructuring and rationalizing its administrative structure, internal cost-cutting measures and cutting staff; (b) introducing more competitive procurement policies; (c) expanding its client base; (d) initiating and supporting technology retention and development programmes; and (e) pursuing export markets through international marketing and marketing support and negotiating offset or counter-trade agreements with foreign suppliers.[4]

Most of Armscor's adjustment strategies which were implemented after 1989 were geared towards the realization of three interrelated goals—to ensure Armscor's survival as the key actor in the domestic military–industrial complex; to ensure the survival of the domestic arms industry, albeit in a slightly

---

[2] *Top Companies* (Financial Mail: Johannesburg, 22 June 1990), p. 130.

[3] *Denel Annual Report 1992/93*, p. 10.

[4] On offsets, see chapter 9, section II in this volume; and Udis, B. and Maskus, K. E., 'Offsets as industrial policy: lessons from aerospace', *Defence Economics*, vol. 2, no. 2 (1990), pp. 151–64, quoted in Hartley, K. and Sandler, T., *The Economics of Defence* (Cambridge University Press: Cambridge, 1995), p. 240. Counter-trade is a form of offset.

smaller form, which constituted the rationale for Armscor's corporate existence and its influence within the state; and to obtain better value for its procurement activities.[5]

## Restructuring and rationalization

Armscor's production and research facilities were rationalized, restructured and down-sized between 1989 and 1992, both in response to the defence cuts and the down-sizing of the domestic arms industry generally and in order to make them more 'commercially' oriented. The corporation also implemented a wide range of internal cost-cutting measures which resulted in over 10 000 jobs being lost between 1989 and 1993.

The formation of Denel in 1992 necessitated the further restructuring and rationalization of Armscor's organizational structure and most of its production and research facilities while a large number of personnel were directly transferred to Denel. Only those personnel who were directly involved with the various aspects of procurement remained at Armscor. A number of senior executives were also transferred to similar positions in Denel in order to ensure a high degree of continuity.[6] Armscor's activities were consolidated at Head Office in Pretoria and in two business units—the IMT in Simonstown and the Alkantpan testing range. These measures led to a reduction in the number of middle-management levels within Armscor.

After the completion of restructuring and rationalization, the organizational structure of Armscor comprised a Board of Directors and a Management Board. The Board of Directors reported to the Minister of Defence, was responsible for the strategic and policy framework within which Armscor policy was formulated, and acted as an independent tender board.[7] Despite the changes in the domestic political environment which occurred after 1989, by early 1994 the Armscor Board remained exclusively white, male and Afrikaans, and there was a high degree of convergence of interest and background (language and race) between all members of the Board. The Board of Directors thus continued to represent the formal expression of South Africa's military–industrial complex by reflecting the broader alliance between the state, the military and capital that emerged within the context of Total Strategy during the 1970s and 1980s.

The day-to-day management of Armscor was carried out by a Management Board which consisted of 12 members: the Managing Director, Secretary, the general managers of Armscor's three procurement portfolios (aeronautics, vehicles and weapon systems, and electronics and maritime) and general managers of public relations, finance and administration, quality control, planning, personnel, foreign procurement and import/export control (see appendix 2).

[5] Private communications with Armscor officials.

[6] For example, Johan Adler, the former Director of Finances and Planning at Armscor, was appointed as Managing Director of Denel in Apr. 1992.

[7] In practice Armscor had more regular contact with the Deputy Minister of Defence than the Minister of Defence after 1992, and it was through the Deputy Minister that Armscor retained its access to the executive levels of the state.

Despite considerable reorganization, there was also a remarkable degree of continuity within the executive and management structures of Armscor after 1992. The majority of senior personnel were white, male Afrikaners who had all had long careers in the domestic arms industry. After 1992 only two outside (non-white) appointments (A. Omar and J. Kgare) were made to the Management Board. According to some commentators, both these were 'soft appointments' in that they did not have any influence with regard to procurement policy or the industry functions, which is where the real power in Armscor was located.[8]

## Competitive procurement policies

A significant aspect of Armscor's adjustment strategy after 1989 was the introduction of a new, more competitive procurement policy which emphasized value for money, competition for contracts and fixed-price rather than cost-plus contracts.[9] The policy of preferential purchasing in favour of the domestic industry was also slowly abandoned after 1990, despite the UN arms embargo. Examples of foreign procurement deals included the purchase of a replenishment vessel, the SAS *Outeniqua*, from Russia in 1993 for $13.6 million to replace the navy's supply ship SAS *Tafelberg*. It was estimated at the time that it would have cost $40 million to build a local replacement for SAS *Tafelberg*.[10] Armscor also concluded a 520 million-rand contract with Pilatus (Switzerland) in 1993 to supply the air force with 60 PC-7 trainer aircraft to replace the Harvard trainers.[11] The awarding of the contract to a foreign company in preference to the local product (the Ovid trainer developed by Denel and Aerotek) was a highly visible example of the shift in Armscor's procurement policy away from preferential purchasing from the domestic industry. The decision to award the contract to Pilatus on the basis of narrowly defined 'value-for-money' considerations was severely criticized at the time by the local arms industry.[12]

## An expanded client base

The dramatic decline in procurement expenditure by the SADF after 1989 forced Armscor to broaden its client base. After 1990 the share of its procurement cash-flow from clients such as the South African Police and Department of Correctional Services increased quite significantly as the share of procurement cash-flow from the SADF declined (see table 5.1).

Armscor's attempt to broaden its client base was part of a broader adjustment strategy aimed at expanding and redefining the corporation's role and range of activities. Armscor offered to make its procurement services available to other public organizations and to the defence forces of other African states. It also

[8] Private communication with Dr R. Williams, Military Research Group, Johannesburg, Apr. 1994.
[9] Private communication with Armscor official, Apr. 1994.
[10] 'Arctic ship goes south', *Jane's Defence Weekly*, 27 Feb. 1993, p. 15.
[11] 'Price, risk determine Ovid deal says SAAF', *Engineering News*, 5 Feb. 1993, p. 1.
[12] 'Pilatus–Ovid dog-fight continues', *Engineering News*, 12 Feb. 1993, p. 1.

**Table 5.1.** Armscor procurement expenditure, 1990–94

|  | 1990 | 1991 | 1992 | 1993 | 1994 |
|---|---|---|---|---|---|
| Total Armscor procurement expenditure[a] (m. rand, constant 1990 prices) | 5 504 | 4 202 | 3 243 | 3 162 | 2 427 |
| % change |  | – 23.7 | – 22.8 | – 2.5 | – 23.2 |
| SADF procurement expenditure (m. rand, constant 1990 prices) | 5 414 | 4 051 | 2 892 | 2 938 | 2 225 |
| SADF as share of total procurement (%) | 98.4 | 96.4 | 89.2 | 92.9 | 91.7 |
| Police and other procurement expenditure (m. rand, constant 1990 prices) | 90 | 151 | 351 | 224 | 202 |
| Police and others as share of total procurement (%) | 1.6 | 3.6 | 10.8 | 7.1 | 8.3 |

*a* Includes procurement for the SADF, the police and other government departments.

*Source: Armscor Annual Report*, various years.

suggested in its annual report for 1993/94 that it was willing to act as the state's general procurement agency, thereby avoiding unnecessary duplication of resources and streamlining the procurement of goods and services for all government departments.[13]

*Technology retention and development programmes*

One of Armscor's most important adjustment strategies after 1989 was support for technology retention and development programmes, particularly in the context of declining defence expenditure. These programmes, which were funded from the General Support Programme and the Special Defence Account of the defence budget, aimed to ensure the long-term survival of the domestic arms industry.[14] In 1991 Armscor established an Industry and Technology Survival Plan (ITSP) in conjunction with the SADF. The ITSP identified and supported critical skills, design capacities and manufacturing processes which were needed for the future requirements of the defence force and which would enable the arms industry to survive and adjust to lower defence expenditure.[15] Armscor also established a Technology Development Programme (TDP) in 1992, which was funded by the DRDC, to develop the skills and technological and manufacturing capabilities needed to design, develop, maintain and manufacture new products (and processes) for future weapon systems.[16] The TDP and ITSP programmes covered the entire spectrum of equipment and *matériel* used by the SADF, and over 300 technology programmes were initiated in 1992 and 1993.[17]

---

[13] *Armscor Annual Report 1993/94*, p. 6.

[14] Private communication with Dr A. Buys, General Manager, Planning, Armscor, Apr. 1994.

[15] The ITSP received 208 million rand in 1991, 260 million rand in 1992 and 105 million rand in 1993 from the defence budget.

[16] *Armscor Annual Report 1992/93*, p. 10.

[17] For a list of the technology programmes which were initiated during 1992 and 1993, see *Armscor Annual Report 1992/93* (note 16) and *Armscor Annual Report 1993/94* (note 13).

Despite Armscor's efforts to ensure the survival of critical defence skills and technologies, military R&D expenditure declined by 65 per cent in real terms between 1989 and 1993 and from 48 per cent of total South African R&D expenditure in 1989 to 18.7 per cent in 1993 (see table 4.5).

Armscor, in conjunction with the various arms of the SADF, also used the Special Defence Account (the SADF procurement budget) to support critical defence skills and technologies within the domestic arms industry, while at the same time attempting to fill some of the SADF's equipment 'gaps'. An increasing share of Armscor's procurement expenditure was directed towards aerospace after 1989, the area in which the SADF had a number of significant equipment gaps. The major aerospace projects carried out between 1989 and 1993 included: the upgrading of Mirage III fighters to Cheetah E configuration in a joint programme with IAI/Elta in Israel; the purchase of the Pilatus PC-7 from Switzerland; the conversion of Puma helicopters to the Oryx configuration; the production of an engineering model of the Rooivalk attack helicopter; and the upgrading of the C-47s for utility, transport and coastal patrol tasks.[18]

## Arms exports and offset agreements

Increasing exports by both value and share of the international arms market was one of the major offensive adjustment strategies which Armscor pursued after 1989. Under its mandate Armscor was responsible for the marketing of South African armaments and also issued export permits. The corporation launched a major advertising strategy in 1992 to increase international awareness of South Africa's arms industry and its products. It also provided marketing support to Denel and to approximately 800 private-sector defence firms to enable them to market their products overseas and to attend international arms exhibitions such as IDEX '93 in Abu Dhabi and the 1993 Dubai Air Show. Armscor arranged the first local arms exhibition, DEXSA, in Johannesburg in November 1992, in which more than 150 local companies participated,[19] and set up five overseas offices to assist with arms imports and exports.[20]

Armscor's advertising strategy, South Africa's participation at international defence exhibitions and the improvement in its foreign relations as a result of the moves to abolish apartheid combined to have a significant impact on South Africa's arms trade (see table 5.2).

South Africa's arms exports increased quite dramatically after 1990 while the value of its arms imports declined, partly as a result of the reduction in the procurement budget. The arms trade balance registered a surplus for the first time in 1993. Another important aspect of Armscor's adjustment strategy which was linked to the promotion of arms exports was the negotiation of offset agreements with foreign suppliers.

[18] For details of procurement projects, see *Armscor Annual Report 1992/93* (note 16); and *Armscor Annual Report 1993/94* (note 13).

[19] *Armscor Annual Report 1992/93* (note 16), p. 15.

[20] Armscor's overseas offices were in France, Israel, Russia, Switzerland and the United Arab Emirates. *Armscor Annual Report 1993/94* (note 13).

**Table 5.2.** The South African arms trade, 1989–94
Figures are in m. rand in constant 1990 prices.

|  | 1989 | 1990 | 1991 | 1992 | 1993 | 1994 |
|---|---|---|---|---|---|---|
| Arms imports | 2 691 | 1 982 | 847 | 618 | 644 | 421 |
| Arms exports | 236 | 163 | 700 | 382 | 625 | 550 |
| Trade balance | − 2 455 | − 1 819 | − 147 | − 236 | − 19 | 129 |
| Counter-trade credits | . . | . . | 131 | 161 | 66 | 245 |
| Overall balance | − 2 455 | − 1 819 | − 16 | − 75 | 47 | 374 |

*Sources:* Private communications with Armscor officials, Apr. 1994; and *Armscor Annual Report*, various years.


Armscor's Board in the early 1990s approved a policy which required an offset or counter-trade of at least 50 per cent of contract value to be negotiated for all foreign contracts worth over 5 million rand.[21] Armscor negotiated seven such agreements between 1991 and 1993 in the attempt to conserve foreign exchange and to support the local arms industry, to a total value in 1993 prices of 2.5 billion rand. If the value of offsets is included, South Africa's overall balance in armaments registered a surplus in 1993 and 1994 (see table 5.2).

## Denel's adjustment strategies, 1992–93

Denel was established as a state-owned industrial company under the Ministry of Public Enterprises in April 1992. It inherited most of Armscor's production and research facilities, assets valued (at book value) at 4.5 billion rand, over 15 000 employees and a share of Armscor's long-term liabilities.[22]

From its inception Denel was faced with a declining domestic defence market and therefore immediately began to pursue a number of adjustment strategies in order to 'commercialize' the former Armscor subsidiaries and to reduce its dependence upon the domestic defence market. Most of Denel's divisions and business units which were former Armscor subsidiaries or research facilities, and which had been established without commercial considerations in mind, were highly dependent upon domestic defence sales and therefore particularly vulnerable to cuts in defence expenditure. According to a study by Bitzinger, a firm's size and degree of dependence on defence business determine its vulnerability to defence cuts.[23] In terms of the matrix developed by Bitzinger, Denel as a company exhibited a medium vulnerability to defence cuts at its creation in 1992, while most of its divisions exhibited a medium or high vulnerability to defence cuts (see table 5.3).

[21] *Armscor Annual Report 1993/94* (note 13), pp. 22–23. On offsets, see note 4.

[22] Denel inherited a proportion of Armscor's long-term liabilities valued at 210 million rand at its inception in Apr. 1992. *Armscor Annual Report 1992/93* (note 16), p. 29; and *Denel Annual Report 1993/94*, p. 51.

[23] Bitzinger, R., *Adjusting to the Drawdown: The Transition in the Defence Industry* (Defence Budget Project: Washington, DC, 1993).

**Table 5.3.** Denel: vulnerability to defence cuts, 1992

| Size/ Turnover | Dependency on defence sales (share of turnover) | | |
| --- | --- | --- | --- |
| | > 80% | 20–80% | < 20% |
| Large > 100 million rand | *Medium vulnerability* LIW Kentron Naschem Simera Infoplan Houwteq | *Medium vulnerability* PMP Somchem Swartklip | *Low vulnerability* |
| Medium 50–100 million rand | *High vulnerability* Eloptro | *Medium vulnerability* | *Low vulnerability* |
| Small < 50 million rand | *High vulnerability* OTR Mexa Gennan Dinmar Advena | *High vulnerability* Mechem Musgrave | *Low vulnerability* Bonaero Park Denprop Denel Insurance Mediamakers |

*Sources:* Adapted from Bitzinger, R., *Adjusting to the Drawdown: The Transition in the Defence Industry* (Defence Budget Project: Washington, DC, 1993); and private communications with Denel officials.

Denel pursued a combination of defensive and offensive adjustment strategies after 1992. Defensive strategies included the laying off of staff, internal cost cutting, closure of certain production facilities, diversification and conversion. Over 1600 personnel were cut in 1992–93, in particular administrative and unskilled staff, while more expensive (white) professional and management personnel were retained. Denel also sold assets, particularly properties and capital equipment, and attempted to down-size and rationalize a number of its weaker divisions, particularly in the Aerospace, Informatics, and Engineering Services groups. Offensive adjustment strategies were aimed at maintaining or increasing the company's defence business (both local and foreign) and included rationalizing and consolidating the defence operations, mergers, acquisitions and joint ventures with local and foreign defence firms, and increasing exports. The offensive and defensive adjustment strategies were not mutually exclusive in that Denel made no attempt to get out of the defence market altogether. Instead it pursued a dual-track approach, adapting to the declining domestic defence market by rationalizing and consolidating its defence operations, while at the same time reducing its dependence on the local defence market by diversifying into civil markets and products and increasing arms exports.

*Restructuring, rationalization and consolidation*

At its creation Denel began to restructure and commercialize the former Arms-cor production and R&D facilities. Denel's business activities were divided into five industrial groups—Systems, Manufacturing, Aerospace, Informatics and Properties, and Engineering Services—which in turn consisted of a number of divisions and business units. Most of Armscor's major arms production sub-sidiaries were consolidated in the Systems, Manufacturing and Aerospace groups, which in turn formed the core of Denel's defence operations. The Informatics and Properties Group and the Engineering Services Group also included a number of former Armscor subsidiaries but were highly geared towards the civil market from the time of their formation.

The organizational structure of Denel was on fairly conventional lines and consisted of a Board of Directors and a Management Board. A large percentage of Denel's senior management appointed after 1992 were former Armscor personnel, reflecting the high degree of continuity and strong institutional and personal links between the two organizations. However, certain tensions began to develop between Armscor and Denel after 1992 in that Denel was forced to compete on equal terms with other (private) defence firms under Armscor's new competitive procurement policy.[24]

In 1993 Denel continued to rationalize its operations and restructured its 18 divisions and subsidiaries (down from 21 in 1992) into six different industrial groups: Systems, Manufacturing, Aerospace, Informatics, Properties and Engineering. The major organizational changes took place in the Informatics and Engineering groups. The Informatics and Properties Group was split into two. The divisions and business units of the Engineering Group, such as Mexa and Gennan, were rationalized during 1993 into one division, Dendex. It concentrated on systems analysis and systems engineering. (See appendix 3 for details of the changing organizational structure of Denel after 1992.)

The strategy of rationalization and restructuring after 1992 was based on a desire to remain in the defence market but to 'shrink to fit' to the declining domestic defence market by consolidating and rationalizing the defence oper-ations. This involved cutting internal costs, for instance, by reducing capital expenditure or laying off workers, while at the same time preserving core capabilities and key operations in defence production. Over 1600 jobs were cut in 1992–93, many of these employees being retained on short-term contracts for specific projects.[25]

*Exports and international joint ventures*

One of Denel's primary offensive adjustment strategies was to increase exports of both military and commercial products. For Denel as well as for Armscor, the improvement in South Africa's foreign relations after 1990 helped it to gain

---

[24] Private communication with T. de Waal, Executive General Manager, Armscor, Apr. 1994.
[25] *Denel Annual Report 1993/94* (note 22).

**Table 5.4.** Denel arms exports, 1992–93

|  | 1992 | 1993 | % change 1992–93 |
|---|---|---|---|
| Total arms exports by Denel | 353 | 406 | *15.0* |
| (m. rand, constant 1990 prices) | | | |
| Systems Group exports | 232 | 271 | *16.8* |
| (m. rand, constant 1990 prices) | | | |
| as share of total (%) | *66* | *67* | |
| Manufacturing Group exports | 121 | 135 | *11.5* |
| (m. rand, constant 1990 prices) | | | |
| as share of total (%) | *34* | *33* | |
| Total South African arms exports | 382 | 625 | *63.6* |
| (m. rand, constant 1990 prices) | | | |
| Denel arms exports as share of total SA arms exports (%) | *92* | *65* | |

*Sources: Denel Annual Report*, various years; and *Armscor Annual Report*, various years.

access to new international markets. With the support of Armscor, Denel began to market its products overseas and attended several international defence exhibitions. A number of its products, such as the G5 and G6 howitzers, are regarded as highly competitive, and during the early 1990s Denel won a number of export orders in the Middle East.[26]

Denel's arms exports increased quite significantly in 1992–93 (see table 5.4). The company exported military equipment to 37 countries in 1992 and 41 countries in 1993 and established a network of agents in 37 countries.[27] The Systems Group, and particularly LIW, which manufactures the G5 and G6 howitzers, provided the bulk of Denel's military export earnings in 1992 and 1993, and both the Systems and the Manufacturing groups witnessed significant real increases in export sales. A significant, although declining, proportion of South Africa's arms export earnings in 1992–93 was Denel's—over 90 per cent of total arms exports in 1992 and nearly 65 per cent in 1993.

In addition to pursuing export markets, Denel entered into a number of joint ventures and strategic alliances with defence and defence-related firms in other countries, for example, cooperation agreements with China North Industries in Beijing and Chartered Industries of Singapore in 1992 and a licensing agreement with Alvis in the UK in 1993.[28] However, the UN arms embargo severely constrained its attempts to enter into formal arrangements with foreign firms.

*Diversification*

All the Denel divisions pursued strategies of diversification in order to reduce their dependence on the local defence market (see appendix 3). These strategies

[26] Iraq, Oman, Qatar and the United Arab Emirates all purchased G5 or G6 howitzers from Denel during the early 1990s. *Jane's Defence Weekly*, 14 Nov. 1992, p. 33; and 13 Feb. 1993, p. 49.
[27] *Denel Annual Report 1993/94* (note 22), p. 11.
[28] 'South African industry targets Asia', *Defence News*, 28 Sep. 1992, p. 34; and *Denel Annual Report 1993/94* (note 22).

**Table 5.5.** Denel: turnover by source, 1992–93

Figures are in m. rand in constant 1990 prices. Figures in italics are percentages.

|  | 1992 | % of turnover | 1993 | % of turnover | % change |
|---|---|---|---|---|---|
| Domestic arms sales | 1 391 | *63* | 1 069 | *53* | *– 23* |
| Arms exports | 353 | *16* | 406 | *20* | *15* |
| Domestic civilian sales | 441 | *20* | 493 | *24* | *12* |
| Civilian export sales | 23 | *1* | 49 | *3* | *111* |
| Total turnover | 2 208 | *100* | 2 017 | *100* | *– 8* |

*Source: Denel Annual Report*, various years.

included joint ventures, acquisitions and/or mergers with civilian firms, the pur-chase of existing non-military product lines or licensing agreements, and the development of commercial, civilian products using defence technology and production facilities (that is, spin-off). They were accompanied by significant investments in R&D and new product development as well as by a major marketing strategy to identify new (local and foreign) markets.[29]

The Systems Group, which was made up of most of the major former Armscor prime contractors, made significant attempts to diversify after 1992. Some divisions, such as Eloptro and Musgrave, diversified by developing civilian products using existing defence technologies. Kentron diversified by acquiring a civilian firm, Irenco, which handled all Kentron's civilian business, and by developing civilian products using its existing defence technology. LIW attempted to diversify through joint ventures with civilian companies and entered into a joint venture with Bell (Pty) Ltd to produce skid-steer loaders, and with Dendex to produce tractors. Mechem was an exception in that it remained highly dependent upon its defence business because of its dominance in the local mine-protected vehicle market.

The Manufacturing Group, which included all the former Armscor ammu-nition manufacturers as well as the Advena Research Laboratories, was highly dependent on the local defence market at the time of the formation of Denel. This group therefore made a concerted effort to diversify and reduce its depen-dence on the local defence market. Most divisions diversified through spin-offs. The group's expertise in ammunition and explosive products meant that most of the divisions developed commercial explosive products for the mining and con-struction industries. Swartklip, for instance, developed a product called the Boulder Breaker. PMP developed a number of brass and pressed products for the automotive industry based on its ammunition production technologies. Some divisions in the group entered into joint ventures with civilian companies: Somchem, for example, entered into a licensing agreement with Tubi Sarplast in Italy to manufacture glass-reinforced polyester pipes.

---

[29] Spending on R&D, which was financed from internal funds, increased by nearly 190% in real terms from 26 million rand in 1992 (15.2% of operating profit) to 81 million rand in 1993 (38.6% of operating profit). Information obtained from private communications with Denel officials, Mar. 1993.

The Aerospace Group, particularly Simera, was one of the major players in South Africa's military aerospace industry at the time of the formation of Denel in 1992. However, after 1992 the group made significant attempts to diversify. Houwteq, which was formerly involved in military satellites, converted all its facilities to civilian purposes and became involved in the development and marketing of low earth orbit (LEO) satellites.[30] Of Houwteq's turnover in 1992 and 1993, 100 per cent was commercial business. Simera, despite being heavily involved in local defence business as the main contractor for the South African Air Force, actively pursued commercial products and markets after 1992 and in 1993 was awarded a contract by Rolls Royce (UK) to supply gearboxes for commercial aircraft engines. Simera also increased its commercial business by winning contracts for the servicing and maintenance of commercial airlines. Overberg Test Range, formerly a test range for the aerospace and missile industry and a launch site for LEO satellites, struggled to reduce its dependence on domestic defence business.

The Informatics and Properties Group, which was split into two separate groups in 1993, actively pursued commercial products and markets in 1992–93. The Informatics Group, formerly involved in military software and information technology, was restructured during 1993 into a number of business units such as Excelsa and Intersolve in order to develop new commercial business. The synergies between military and commercial software and information technology and its significant human resources placed the group in a relatively competitive position with respect to South Africa's information technology sector. Unlike the other groups in Denel, it experienced a real increase in turnover in 1992–93, despite a very competitive domestic market for information technology products and services. The Properties Group, which was involved in commercial property development and management, experienced a real increase in turnover in 1993 despite the depressed conditions of the local property market (see appendix 3).

The aim of rationalizing the Engineering Group into one division (Dendex) in 1993 was to enhance the commercial business of the group. The Gerotek Test Facility, which was formerly used exclusively for military testing, started to attract significant commercial business and in 1993 was awarded a contract for tyre tests by Continental Tyres (Germany).

Only one of Denel's divisions, Houwteq, pursued a strategy of conversion which involved the transformation of all its resources and productive capacities from military use to civilian purposes. The issue of conversion is examined in more detail in chapter 8.

[30] Private communication with F. Herbst, General Manager, Houwteq, July 1993.

**Table 5.6.** Denel groups: defence share in turnover, 1992–93
Figures are percentages.

| Group | Defence as share of turnover 1992 | Defence as share of turnover 1993 | Change in turnover 1992–93 |
|---|---|---|---|
| Systems | 61 | 53 | – 1.3 |
| Manufacturing | 44 | 28 | – 9.0 |
| Aerospace | 50 | 78 | – 19.8 |
| Informatics | 85 | 80 | 9.8 |
| Engineering | 79 | 82 | – 28.5 |
| Properties | 0 | 0 | 46.6 |
| **Denel total** | **63** | **53** | – 8.7 |

*Source: Denel Annual Report*, various years.

## Denel: adjustment experience, 1992–93

The outcome of Denel's adjustment in 1992 and 1993 was reflected in the changing composition of the company's business as measured by turnover. Its domestic defence business declined quite dramatically both in real terms and as a share of turnover. This was primarily a function of the 23 per cent real decline in SADF procurement expenditure in 1992–93. However, the declining contribution of domestic defence business was offset to some extent by the real increases in the value of exports and commercial business, with the result that the net fall in turnover was limited to 8 per cent. The significant real increase in exports, both in absolute terms and as a share of turnover, reflected the success of South Africa's arms export drive. The bulk of Denel's export earnings were accounted for by the Systems Group (particularly LIW and Kentron) and the Manufacturing Group (particularly Naschem and PMP).[31] The increase in the value and share of Denel's civilian business reflected the company's concerted efforts to diversify and develop new civilian products and markets. The bulk of its new civilian business in 1992–93 was accounted for by the Informatics and Property groups, and also increasingly by the Manufacturing Group (see table 5.6).

The contribution of the various groups to Denel's turnover also changed in 1992–93 as a result of the changing nature of the company's business. Some groups struggled to reduce their dependence upon the local defence market (as did Aerospace) and witnessed significant declines in turnover and share, while others such as Informatics were able to find new foreign and local markets for their (military and civilian) products and thus limit, and in some cases offset, the negative effects of the shrinking domestic defence market.

---

[31] The Systems Group accounted for 62% of Denel's exports in 1992 and 61% in 1993, while the Manufacturing Group accounted for 38% of exports in 1992 and 38% in 1993. *Denel Annual Report 1993/94* (note 22).

The Systems, Manufacturing and Aerospace groups tended to have the most difficult and costly adjustment experiences. This was a function of these groups' high degree of dependence on defence business and defence-specific technology, since they had inherited most of Armscor's major production subsidiaries. Both the Systems and the Manufacturing groups saw real reductions in turnover in 1992–93 as a result of real declines in the value and share of their defence business. Both pursued diversification strategies, but found the experience both difficult and costly.[32] Where groups experienced real increases in the value and share of their export and commercial business this was not sufficient to offset the effects of the decline in their defence business. The Aerospace Group, quite surprisingly, witnessed real increases in the value and share of its defence business, partly as a result of a large number of air force procurement projects.[33] However, the increase in domestic defence business was not sufficient to offset the real decline in turnover, which was largely a function of the poor performance of Houwteq, which failed to find local or foreign partners for its satellites.[34] The Engineering Services Group also witnessed a real decline in turnover in 1993 despite an increase in the share of its defence business.

As expected, the only group which witnessed a real increase in turnover in 1993, despite a decline in its defence business, was the Informatics Group. It managed to increase the value and share of its civilian business so significantly that it was able to offset the declining value of its defence business. It was able to adjust relatively successfully because when Denel was formed it was not highly dependent upon defence-led or defence-specific technology: it was involved in software and information technology which was largely civilian-based technology.

As regards the adjustment experiences of Denel's divisions, those which had a high degree of dependence on domestic defence sales and a high degree of dependence on defence-specific technology, such as LIW, tended to have the most difficulty. Those which had a lower dependence on defence-specific technology, such as Infoplan, had easier adjustment experiences. In terms of the financial performance of the various divisions, it is evident that those with a low dependence on defence-specific technology, such as Infoplan, or those with a lower dependence on defence-specific technology and a lower dependence on defence sales, such as PMP, had relatively easy adjustment experiences. These divisions experienced real increases in turnover despite real decreases in the value and share of their domestic defence business. Those with a high degree of dependence both on defence sales and on defence-specific technology, such as LIW and Naschem, had difficult adjustment experiences as evidenced by large real declines in the value and share of their domestic defence business, turnover and employment.

[32] Private communication with M. Koorts, Director, Systems Group, Denel, Mar. 1994.
[33] Private communication with B. Kruger, Director, Aerospace Group, Denel, Mar. 1994.
[34] Private communication with F. Herbst, General Manager, Houwteq, Apr. 1994.

## Nuclear weapon industry adjustment strategies

South Africa's nuclear arms industry, which was located at the Atomic Energy Corporation (AEC) at Pelindaba and Advena, Armscor's central laboratories, was dismantled after 1989 with the termination of the nuclear weapon programme. The facilities which had been used in the manufacture of nuclear material and the construction of nuclear weapons were decontaminated and converted to civilian production. The commercialization of Armscor's production facilities and the formation of Denel in 1992 provided a model for the commercialization and conversion of South Africa's nuclear weapon industry.

The AEC's pilot uranium enrichment plant at Pelindaba, which manufactured the material for nuclear weapons, was decommissioned in early 1990. Thereafter the AEC began a process of commercializing its operations and facilities. It grouped together facilities and technologies with commercial potential into 11 business units, which in turn were grouped together under a new state-owned, commercial management unit called Pelindaba Technology Products (PTP) in April 1993.

Each business unit of PTP was expected to be independently profitable, and all technology development costs, product industrialization costs, capital costs and initial start-up costs were to be borne by the individual business units.[35] The business units drew on expertise and technologies in four key areas: radio-isotopes, fluorochemicals, engineering systems and mechanical products which had been developed as a result of the uranium enrichment programme.[36]

PTP pursued similar adjustment strategies to those of Denel through the application, exploitation and conversion of existing skills, technologies and products for commercial purposes. This diversification and conversion process was achieved through increased marketing efforts (domestic and foreign), the pursuit of new domestic and export markets, and joint ventures.[37] Some of the business units, such as Flosep and Apogee, also established marketing offices in foreign countries. According to Karl Voigt, the Executive General Manager of PTP, the major problem of commercialization involved 'retraining staff to think along business lines . . . previously it was a case of budgeting for expenditure and simply meeting that budget . . . now we have to think of generating income to satisfy shareholder [i.e., the state's] expectations'.[38] During its first year of operation, PTP generated a turnover of 85 million rand (2–3 per cent derived from exports), which was sufficient to cover its current operating costs.[39]

The occupation of new, expanded facilities at Advena was under way when the nuclear weapon programme was terminated. In April 1992 Advena's staff

---

[35] *SALVO*, vol. 2 (1993), p. 20.
[36] *Business Day*, 7 Oct. 1993, p. 10.
[37] PTP signed a joint venture with Fedgas to provide compressed fluorine for use in plastic products, and with Unifill to provide air filters for the motor industry.
[38] Karl Voigt, Executive General Manager, PTP, quoted in *Business Day*, 7 Oct. 1993, p. 10.
[39] See note 38.

and facilities became part of the Manufacturing Group of Denel, and in April 1993 Advena ceased to be a separate division of Denel and its business activities, including personnel, were integrated with another division, PMP, in an attempt to cut overhead costs. Advena's adjustment strategies included staff cuts (over 200 from 1989 to 1993), joint ventures and the application of existing skills and technologies to commercial products.[40] The development of commercial products was based on Advena's capabilities in high explosives, pyrotechnics, metallurgy, high-speed electronics, environmental and reliability testing, and ultra-high-speed diagnostics which had been developed as a result of the nuclear weapon programme.[41]

Advena's adjustment efforts were constrained by the severe domestic recession and by the fact that the 'high overhead costs associated with fixed assets and a highly skilled (and expensive) workforce, customized to the (nuclear) weapons programme were not conducive to a competitive price structure needed for rapid commercial market penetration . . . The mindset of scientific and technical staff had to be converted from being technology driven to being market driven and business oriented'.[42]

## Adjustment in the public-sector arms industry: an assessment

Assessing the adjustment experience of the public-sector arms industry between 1989 and April 1994, it is evident that Armscor managed to adjust to the new conditions of the domestic defence market quite successfully. Denel, on the other hand, found its adjustment experience more challenging, especially in the context of the severe domestic recession and a highly competitive international arms market.

With the formation of Denel in 1992, Armscor was able to concentrate on its primary function, procurement, while at the same time trying to redefine and expand its role within the context of South Africa's changing strategic, political and economic environment. Furthermore, the formation of Denel at arm's length from Armscor allowed the latter to introduce more competitive procurement policies, whereas previously it had been forced to procure from its own subsidiaries. While Denel remained a prime contractor for Armscor after 1992, given the numerous informal and formal links between the two organizations, a number of private companies were appointed as main contractors in preference to Denel companies such as Atlas Aviation and Kentron. As part of this new competitive procurement policy, Armscor also began to procure weapon systems from abroad in preference to local products (as in the case of the Pilatus contract for trainers for the South African Air Force).

---

[40] Advena in conjunction with private-sector firms (e.g., Trinity Trigate) developed a medical ventilator. A contract was signed with the Transvaal Provincial Administration in 1993 to provide all hospitals in the Transvaal with ventilators.

[41] SALVO, vol. 2 (1993), p. 21.

[42] G. Enslin, General Manager, Advena, quoted in Albright, D., 'The legacy of the South African nuclear weapons program', Paper presented at the Conference on a Nuclear Policy for a Democratic South Africa, Cape Town, 11–13 Feb. 1994, p. 12.

Armscor's attempts to increase South African arms exports proved relatively successful, as evidenced by the significant increases in the country's foreign arms sales after 1990. Armscor was also particularly astute in continuing to highlight the purported economic benefits of domestic arms production and arms exports (i.e., foreign exchange earnings and job retention) in an attempt to counter the growing domestic and international opposition to South Africa's increasing involvement in the international arms trade.

Most importantly, despite being relieved of its arms production activities, Armscor managed to maintain its position as the centrepiece and overall coordinator of the domestic arms industry, including control over arms imports and exports. It was also able to maintain certain critical technologies and industrial capabilities within the domestic arms industry despite the drastic cuts in procurement expenditure. Furthermore, it managed to maintain its institutional power and influence within the state and the Department of Defence while at the same time attempting to redefine and even enhance its role. It was therefore able to ensure that it would survive into the post-apartheid era with its institutional and ideological interests intact.

Denel found the adjustment experiences of its first two years of existence particularly difficult. The company inherited a large, highly skilled, specialized and highly paid workforce which had been relatively isolated from commercial market forces and a highly specialized, expensive production infrastructure which was not geared towards mass commercial production. Furthermore, the company came into existence in the context of a severe domestic recession, an already depressed commercial market and an increasingly competitive international defence market. Despite these significant technological, organizational, cultural and economic barriers to entry into both commercial and defence markets, the company remained profitable in 1992–93. Although it experienced a real decline in turnover in 1993, it declared a dividend to its sole shareholder, the government, in both 1992 and 1993.

Denel's continued financial viability after 1992 was achieved largely through massive job cuts, internal cost cutting and the sale of assets (e.g., property) and tended to obscure a number of fundamental weaknesses in the company, such as high overheads, large surplus capacities, duplicated facilities in many divisions, a reliance on the state for R&D expenditure, a declining return on assets, and heavy dependence on a rapidly shrinking domestic defence market.

The fall in Denel's share of the local military market—by over 23 per cent in real terms in 1992–93—was partly a function of the real reduction in SADF procurement spending but was also in part a function of Armscor's more competitive procurement policies and of the fact that Denel was increasingly 'losing out' either to local private-sector contractors (which had previously been subcontractors to the Denel companies) or to foreign defence contractors.[43] This further weakened its financial position. In some cases its

---

[43] *Armscor Annual Report 1992/93* (note 16) and *1993/94* (note 13); and *Denel Annual Report, 1992/93* and *1993/94* (note 22).

facilities were duplicated in the private sector, which tended to be more cost-effective and flexible, and this made it more difficult for Denel to offer competitive tenders.

Denel adopted a combination of defensive and offensive adjustment strategies in response to the declining domestic demand for armaments. The most successful were increasing exports and diversification. The relative success of these strategies was reflected in the growing number of civilian products and joint ventures with local and foreign civilian firms as well as in the increasing share of exports and civilian sales as a percentage of turnover.

In interviews with personnel from the private-sector arms industry it became clear that the absence of a clearly defined institutional relationship between Armscor and Denel after 1992 caused widespread confusion and resentment within certain parts of the private-sector arms industry.[44] Many private-sector defence firms believed that Armscor had divided loyalties and that Denel received preferential treatment in the awarding of defence contracts because of the informal (personal) linkages that existed between them. Many private-sector firms also felt that Denel had an unfair advantage in competing for defence contracts, in that it had inherited its assets from the state cost-free and could therefore afford to undercut prices in the private sector. Another criticism levelled at Denel was that it began to pursue a strategy of vertical integration, contracting out less work to subcontractors in the private-sector arms industry. Thus after 1992 Denel started to compete in markets which were previously the exclusive domain of the private sector—electronics, avionics and information systems)—thereby undermining the government's stated policy of not duplicating facilities and capabilities that already existed in the private sector.

## III. Defence industrial adjustment in the private-sector arms industry

The defence cuts and disarmament measures which led to the restructuring and commercialization of South Africa's public-sector arms industry had a dramatic effect on the country's private-sector arms industry. The changes in Armscor's procurement policies, which entailed more competitive procurement, as well as procurement from abroad, together with the formation of Denel as a contractor and competitor, fundamentally altered the 'cosy' relationship that had been built up between the public-sector arms industry and the private sector since the early 1960s. The cuts in procurement expenditure, which led to the cancellation and/or postponement of major weapon projects, resulted in a dramatic downsizing of the private-sector arms industry and were accompanied by declining output, reduced profitability, large-scale job cuts and the exit of a number of firms from the defence market. In this context private-sector defence firms were forced to pursue a number of supply-side adjustment strategies in order to

[44] Private communication with Gen. (ret.) J. Kriel, Executive Director, South African Defence Industry Association (SADIA), Apr. 1996.

**Table 5.7.** Structure of the private-sector arms industry, 1993

| Defence business as share of turnover | > 1 | > 10 | > 20 | > 30 | > 40 | > 50 | > 60 | > 70 | > 80 | >90 | 100 |
|---|---|---|---|---|---|---|---|---|---|---|---|
| Number of firms or business units | 1 468 | 769 | 398 | 188 | 38 | 40 | 18 | 16 | 22 | 12 | 4 |

*Source:* Private communication with Dr A. Buys, Armscor, Apr. 1994.

minimize the impact of the procurement cuts. In considering their responses to defence cuts most private-sector firms were motivated by both survival and profitability, for in the short run they were constrained by their existing plant, labour, locations, markets and distribution systems. This section examines the adjustment experiences of South Africa's private-sector arms industry between late 1989 and April 1994.

**The size and structure of the private-sector arms industry**

According to information obtained from Armscor, nearly 3000 private-sector firms and business units were directly involved in domestic arms production in the early 1990s, whether as contractors, subcontractors or suppliers (see table 5.7).

In 1993 private-sector defence firms (2973) represented 14 per cent of all firms in the manufacturing sector.[45] In terms of the orientation of the firms, only four were dependent on defence for 100 per cent of their turnover, 112 (3.7 per cent) for more than 50 per cent of turnover, and 2237 (75 per cent) for less than 10 per cent of turnover. That such a large number of firms were dependent on defence for less than 10 per cent of their turnover was related to the low degree of vertical integration (i.e., the ability of one firm to act as a supplier on more than one level of the product hierarchy) in the industry.[46] Despite the large number of firms, the private-sector industry was dominated by three large, publicly quoted industrial groups, Altech, Grintek and Reunert, which continued to dominate the private-sector arms industry and certain sectors of the domestic defence market (e.g., electronics) during the period between 1989 and early 1994.

Most private-sector defence firms were concentrated in the PWV region.[47] The domestic defence market consisted of seven sectors—electronics, vehicles, aerospace, ammunition, weapon systems, naval ships and support equipment. Some sectors, such as electronics and naval ships, were dominated by private-sector companies. Altech, Grintek and Reunert operated in most of the sectors of the defence market. There was also a high degree of subcontracting between

---

[45] *South African Statistical Yearbook 1994* (Central Statistics Service: Pretoria, 1994).
[46] Private communications with Armscor officials, Apr. 1994.
[47] See chapter 4, section IV in this volume; and table 4.9.

the various sectors, the result of the low degree of vertical integration in the arms industry.

The ownership patterns of the private-sector arms industry corresponded with the ownership patterns and structures of the manufacturing sector and the economy as a whole. Four of the country's six large financial–mining–industrial conglomerates—Anglo-American, the Rembrandt Group, Anglovaal, Old Mutual, Liberty Life and Sanlam, which between them controlled approximately 90 per cent of the asset value of the Johannesburg Stock Exchange in 1993—owned or controlled one of the large private-sector defence groups.[48] Reunert was controlled by Old Mutual, Grintek was owned by Anglovaal, and Altech was owned by Anglo-American. Other major defence firms such as Plessey Tellumat and Truckmakers were owned by Sanlam. The rest of the private-sector arms industry consisted of small companies listed on the Johannesburg Stock Exchange, such as Spescom and Logtek, in which the majority of shares were owned by individual shareholders or by the directors of the company, or of small and medium-sized unlisted, family-owned companies.

## Private-sector adjustment strategies

The private sector tends to be more sensitive to market forces and conditions than the public sector, and thus most private-sector defence firms anticipated that South Africa's withdrawal from Angola and Namibia in 1989 would result in defence cuts and a decline in domestic demand for armaments. As a result most private-sector firms responded more quickly than the public sector to the defence cuts which were implemented after 1989 and began to adopt a number of supply-side adjustment strategies.

The significant differences between defence firms in the private sector in size, products, position on the product hierarchy, degree of dependence on defence business and ownership meant that a wide variety of adjustment strategies were pursued in response to the changing domestic defence market. Some firms pursued offensive adjustment strategies geared to maintaining or increasing their defence business. These offensive strategies included increasing arms exports, joint ventures with local or foreign defence firms and the monopolization of existing defence markets. Some firms pursued defensive adjustment strategies which involved reducing their defence business or exit from the defence market altogether, cutting staff, selling off defence operations, and diversification and/or conversion. Most firms, however, pursued a combination of defensive and offensive adjustment strategies in order to remain in the defence market but to 'shrink to fit' or adapt to the changing defence market. At no stage between 1989 and early 1994 did the South African Government initiate any public policies, such as manpower policies, science and technology policies, regional policies or industrial policies, in order to help firms to adjust.

[48] Lewis, D., 'Markets, ownership and manufacturing performance', in Joffe, A. *et al., Improving Manufacturing Performance in South Africa* (University of Cape Town Press: Cape Town, 1995), pp. 149–50.

*Exit from the defence market*

One of the most common defensive adjustment strategies for firms with a low exposure to the defence market is to exit the defence market by selling or by closing down their defence operations. In South Africa some private-sector defence firms attempted to leave the defence market after 1989. Dorbyl Marine, the country's major naval shipbuilding firm, closed down its naval shipbuilding facilities in Durban in 1993 because of a lack of defence work. Gencor sold its military vehicles business, Sandock Austral, which manufactured the hulls for the Rooikat armoured car and the Ratel infantry fighting vehicle (IFV) to Reumech in the early 1990s. Sanlam sold Truckmakers, which produced the Samil range of military vehicles, to Reumech in early 1994.

*Down-sizing of defence operations*

Most private-sector defence firms took steps to down-size their defence operations in response to the defence cuts after 1989. This included internal cost cutting, reductions in R&D and capital expenditure, and significant job losses.

Most defence firms cut staff (professional staff and workers) in response to the cuts. The severe recession, which limited the ability of firms to develop new products and markets, increased the pressure on them to do this. Some of the large defence firms, such as Dorbyl and Reutech, cut more than half their total workforces in the period 1989–93. Some large electronics firms such as Altech and Grinaker experienced less dramatic job losses, despite difficult operating conditions and the recession.[49] The scale of job losses was often less drastic among the middle-sized and smaller companies, and some firms, such as XCEL Engineering, actually saw increases in staff numbers either as a result of new defence business or as a result of new commercial markets and products.[50]

In most small and medium-sized defence firms, attempts were made to retain professional staff such as engineers at the expense of clerical staff and semi-skilled and unskilled workers, given the high costs of training and the value-added contribution of professional staff.[51] This corresponded with patterns in the public-sector arms industry. A common feature among many firms (such as ISIS) was a change in hiring practices in favour of contract staff as opposed to full-time permanent staff, usually for specific projects or contracts.[52] In very few firms was retraining for civilian production seen as a priority. In most cases, particularly among smaller companies, it was seen as an expense that companies simply could not afford.[53] At no stage from 1989 to early 1994 did the South African Government initiate any public manpower policies involving

---

[49] Private communications with W. Kotze, Dorbyl Heavy Engineering, Feb. 1994; V. Davis, Reumech, Mar. 1994; and H. van Rensburg, Reutech, Mar. 1994.

[50] Private communication with F. le Roux, XCEL Engineering, Apr. 1994.

[51] Private communication with H. van Rensburg, Reutech, Mar. 1994.

[52] Private communication with C. Neethling, ISIS, Mar. 1994.

[53] Private communication with J. Pougent, Booyco Engineering, Mar. 1994, and G. Peaceful, Comnitech, Feb. 1994.

retraining schemes, job information or encouragement of labour mobility in order to assist displaced defence workers.

Despite reducing defence production, many private-sector firms continued to experience stagnant turnover and declining profitability as reflected in lower share prices, lower dividends and lower earnings per share.[54] However, some of the larger firms, such as Reutech and Reumech, were able to rely on the resources of their parent companies to cross-subsidize or cover the declining profitability of their defence operations. By down-sizing their defence operations, many firms in the private sector began operating at suboptimal levels, with high levels of surplus capacity. The manufacturing sector at an aggregate level saw a decline in the utilization of production capacity from 84 per cent in 1989 to 77 per cent in 1993, the machinery sector a decline from 78 per cent in 1989 to 68 per cent in 1993, and the electronics sector a decline from 76 per cent to 70 per cent over the same period.[55] Most firms made no attempt to retool defence plant and equipment for civilian purposes: it was regarded as too costly and 'a waste of time'.[56] The government also made no attempts to initiate capital policies to assist firms with retooling or investing in new plant and equipment, thereby retaining productive capacity.

*Diversification*

Diversification through the acquisition of civilian products and/or firms was a common adjustment strategy among private-sector defence firms, particularly the larger firms, which had sufficient resources to fund their acquisitions. In most cases firms diversified by acquiring civilian product lines, by acquiring or merging with civilian companies, by entering into joint ventures or through co-production agreements with local or foreign civilian companies. Most of the large defence firms, such as Altech, Grinaker, Reumech and Reutech, were able to rely to some extent on the resources of their larger, more diversified, parent companies in order to fund or facilitate their diversification efforts. The recession in South Africa between 1989 and 1993 also helped some of the larger cash-rich defence firms to acquire certain civilian firms relatively cheaply (an example is Reumech's acquisition of Sandock Austral from Gencor).[57]

Some of the large defence firms pursued diversification through acquisitions quite aggressively. Reutech acquired a number of civilian electronics firms[58] and Reumech acquired a 30 per cent stake in Iveco, an Italian commercial truckmaker, and diversified its commercial business to include the licensed production of refrigerated trailers from Germany and tractors from Italy.[59] Some of the larger firms, such as Dorbyl, which witnessed dramatic declines in

---

[54] Many firms in the electronics sector witnessed declines in their share prices and earnings per share between 1989 and 1993 as a result of the impact of cuts in spending by Telkom, the Post Office and the SADF during the early 1990s. *Top Companies* (Financial Mail: Johannesburg, June 1993), pp. 107–11.
[55] *South African Statistics Yearbook 1994* (note 45).
[56] Private communication with J. Pougent, Booyco Engineering, Mar. 1994.
[57] Private communication with V. Davis, Reumech, Mar. 1994.
[58] Private communication with H. van Rensburg, Reutech, Mar. 1994.
[59] Private communication with V. Davis, Reumech, Mar. 1994.

turnover and profitability after 1989, lacked the resources to diversify out of defence.[60] The smaller defence firms, because of a lack of resources, also found it difficult to diversify through acquisitions of civilian firms.

Spin-off was a common adjustment strategy because it allowed firms to utilize existing workforces, technologies and production processes to develop civilian products. Spin-off strategies were widely used by those defence firms which utilized dual-use technology (such as electronics) or technology which was not highly defence-specific. All the large electronics firms, such as Altech, Grinaker, Plessey and Reutech, used spin-off strategies to develop civilian products. Grinaker Avitronics, which specializes in electronic warfare systems, developed a number of civilian products such as a laser range-finder for use in the mining industry, using its expertise in microwave, electro-optical, digital and software technologies.[61] However, the success of these spin-off strategies, especially for defence electronics firms, was constrained by the continuing domestic recession and by the fact that the civilian electronics markets in South Africa were saturated and dominated by a number of large industrial groups and multinational companies.

Where successful spin-off took place it was determined to a large extent by the resources available to invest in R&D to develop new civilian products, to retrain workers, to retool equipment for commercial, high-volume production, to employ civilian marketing people, and to embark on a commercial marketing strategy.[62] Thus the larger firms (such as Reutech) which had the resources to invest in developing and marketing new civilian products were successful in developing spin-off products. Some of the medium-sized defence firms (e.g., Logtek) were also successful in developing new civilian products and services because they occupied a relatively low position on the product hierarchy.[63] Small defence firms tended to have limited successes with spin-offs, because of problems on the demand side—the recession—and on the supply side—a lack of resources to invest in R&D, retooling and marketing. Generally, the firms in the electronics and military vehicles sectors, which had a lower dependence on defence-led or defence-specific technology, had the most success with spin-off strategies.

Most private-sector firms had some success with diversification, particularly the acquisition of civilian firms or products. However, some small defence firms, such as Northbend and Spesaero, which had a high degree of dependence on defence-led and defence-specific technology, found that diversification was expensive and difficult, that their skills, facilities and products were inappropriate for the civilian market and that they had a better chance of survival by remaining in the defence market.[64] Despite the difficulties associated with diversification, most of the firms remained committed to reducing their defence

[60] Private communications with N. de Villiers, BUSAF (Cape), Feb. 1994 and W. Kotze, Dorbyl Heavy Engineering, Feb. 1994.
[61] Private communication with A. Holloway, Grinaker Avitronics, June 1993.
[62] Private communication with A. Holloway, Grinaker Avitronics, June 1993.
[63] Private communication with B. Cilliers, AMS, Feb. 1994 and N. Moolman, Logtek, Feb. 1994.
[64] Private communication with P. van Rooyen, Spesaero, Feb. 1994.

business, given the prospect of further declines in domestic defence expenditure and the increasing competition from overseas defence suppliers.

As a result of these strategies of diversification most private-sector firms witnessed increases in the value and share of their civilian business. According to information supplied by the South African Defence Industry Association (SADIA), the combined domestic civilian business of the three large defence groups (Altech, Grintek and Reunert) increased by 70 per cent in real terms in 1992–93, and from 5 per cent of total sales to 8 per cent. Exports of civilian goods increased by 36 per cent in real terms and from 5 per cent of total sales to 7 per cent.[65] These trends were reflected in many smaller private-sector defence firms.

*Consolidation of defence operations*

One of the most common offensive adjustment strategies for medium and large firms with a high dependence on defence that wish to remain in defence is to consolidate their defence operations while at the same time preserving core capabilities and key operations. Many firms in the private sector consolidated their defence operations, usually within a separate division or business unit. In some firms defence production was kept physically separate from the firm's civilian business. This adjustment strategy was particularly common in the large defence firms, which were often part of larger industrial companies. Both Reumech and Reutech, which are part of Reunert, rationalized and consolidated their defence business within certain divisions after 1989. In Reumech, defence business which was previously spread across five divisions (Sandock Austral, Ermetek, OMC Engineering, Austral Engineering and Gear Ratio) was rationalized into one division, OMC Engineering.[66] Altech's defence business, which is in six companies (Airconcor, ISIS, Synertech, Teklogic, Telsis and UEC Projects) was consolidated into one division, Altech Systems.[67] The consolidation of defence operations was also common among small and medium-sized defence firms such as XCEL Engineering and AMS.

*Concentration and monopolization of defence markets*

Defence firms that wish to maintain or increase their share of a declining defence market tend to pursue strategies of concentration and monopolization through joint ventures or mergers with and acquisitions of other defence firms. Some firms can increase their share of a particular defence market through specialization, narrowing the range of products produced, and/or aggressive undercutting of competitors. Reumech's acquisition of Truckmakers and Sandock Austral consolidated its already strong position in the military vehicle sector.

[65] South African Defence Industry Association, *Some Statistics of the SADIA Members of the Defence Industry* (SADIA: Johannesburg, 1996).

[66] Private communication with V. Davis, Reumech, Mar. 1994.

[67] Private communication with Dr H. Steyn, Altech, Mar. 1994.

Some firms were able to increase their share of a particular defence market because of the exit of others from the market or because of the absence or weakness of domestic competitors. Small firms such as Aerosud and ATE managed to increase their shares of the domestic aerospace market at the expense of other public-sector competitors (such as Kentron and Simera) and large private-sector competitors (such as Grinaker Electronics) by aggressive cost cutting, smaller overheads and greater flexibility. As a result of these strategies of concentration and monopolization, some firms, such as XCEL, actually increased their dependence on defence after 1989.[68] In some cases firms such as Spesaero, which had previously attempted to diversify out of defence but had found the experience difficult, expensive and a threat to their survival, tried to increase their defence business through the monopolization of defence markets.[69]

As a result of these offensive adjustment strategies, the structure of the domestic defence market changed quite dramatically between 1989 and 1993. The private sector's share actually increased after 1989, and in 1993 accounted for nearly 58 per cent of domestic procurement expenditure, up from 47 per cent in 1992. This increase was achieved at the expense of Denel, whose share fell by 23 per cent in real terms, from over 52 per cent in 1992 to 42 per cent in 1993. The value and share of arms imports remained relatively constant in 1992–93 (see table 5.8).

*Increasing exports*

One of the most common offensive adjustment strategies for defence firms faced with a declining domestic defence market is to increase exports. Most private-sector firms began to pursue export markets aggressively after 1989 in an attempt to offset the declines in the domestic defence market. The improvement in South Africa's international image resulting from the start of negotiations between the ANC and the government, together with Armscor's active international marketing support, helped the private sector gain access to new defence markets while the introduction of government export incentives (such as the GEIS, introduced in 1990) encouraged firms to export their surplus defence products and in some cases reorient their production activities towards foreign defence markets. Some of South Africa's military products, such as artillery systems, high-mobility armoured vehicles, mine-protected vehicles and tactical communications, were internationally competitive, especially in terms of price (helped by a weak currency), and the fact that they were tried and tested during South Africa's war in Angola and Namibia also contributed.[70] Some larger defence companies, such as Altech and Reutech, were able to capitalize on their significant resources and extensive links with multinational companies in order to find new export markets for their defence products.[71]

[68] Private communication with F. le Roux, XCEL Engineering, Apr. 1994.
[69] Private communication with R. van Rooyen, Spesaero, Feb. 1994.
[70] 'Back in the arms bazaar', *Jane's Defence Weekly*, 14 Nov. 1992, pp. 33–45.
[71] Private communications with H. van Rensburg, Reutech, Mar. 1994; and Dr H. Steyn, Altech, Mar. 1994.

**Table 5.8.** Sources of supply for domestic arms procurement, 1992–93

|  | 1992 | 1993 | % change 1992–93 |
|---|---|---|---|
| Total procurement expenditure[a] (m. rand, constant 1990 prices) | 3 243 | 3 162 | – 2.5 |
| Imports (m. rand, constant 1990 prices) | 618 | 644 | 4.2 |
| as share of total procurement exp. (%) | 19.0 | 20.3 | |
| Domestic procurement expenditure (m. rand, constant 1990 prices) | 2 625 | 2 518 | – 4.0 |
| as share of total procurement exp. (%) | 81.0 | 79.3 | |
| Denel domestic arms sales (m. rand, constant 1990 prices) | 1 391 | 1 069 | – 23.1 |
| as share of total domestic procurement exp. (%) | 52.9 | 42.4 | |
| Total private-sector domestic arms sales (m. rand, constant 1990 prices) | 1 234 | 1 449 | 17.4 |
| as share of domestic procurement exp. (%) | 47.1 | 57.6 | |

[a] Includes procurement for the SADF, the police and other government departments.

*Sources: Armscor Annual Report*, various years; *Denel Annual Report*, various years; and other company annual reports, various years.

Some small defence firms (such as Milkor) which produced niche market products were relatively successful in finding new export markets. In some smaller firms earnings from arms exports started to exceed the income from domestic defence sales, and many of these firms began to develop products almost exclusively for the export market (Northbend is an example). All the private-sector firms which were successful in finding export markets for their defence products benefited under category 4 (manufactured products) of the GEIS, which paid up to 19.5 per cent of the total value of exports in tax-free benefits.[72] Some firms, such as Reumech, also begun to pursue export markets in the upgrading and maintenance of weapon systems, particularly in countries in Africa which did not have the skills or the technology to maintain their existing weapon systems, most of which were acquired from foreign suppliers.[73] Most private-sector firms, particularly the larger firms, pursued joint ventures and alliances with foreign defence firms after 1989 in order to capture new markets, spread the costs of new product development and marketing, and attempt to minimize the impact of the declining domestic defence market. However, the UN arms embargo inhibited international joint ventures with foreign defence firms until after the elections in April 1994.

As a result of the pursuit of export markets, the value and share of the private sector's export business (both defence and civilian) witnessed significant increases in the period 1989–93. According to SADIA, the combined value of civilian exports for the three major defence groups (Altech, Grintek and

[72] Private communication with Dr A. Hirsch, School of Economics, University of Cape Town, Mar. 1993.

[73] Private communication with V. Davis, Reumech, Mar. 1994.

Reunert) increased by 36 per cent in real terms, from 5 per cent of total sales to 7 per cent of total sales.[74] The share of the private sector in total arms exports increased from 7 per cent in 1992 to 35 per cent in 1993.[75]

## Adjustment in the private-sector arms industry: an assessment

The private sector's experience of adjustment in the period 1989–94 differed quite substantially from that of the public sector. The private sector is essentially more responsive to market conditions than the public sector, and thus anticipated the defence cuts and the changing strategic and political environment. Especially among the smaller defence firms, it tends to be more flexible and quicker to adapt to changing market conditions, and thus was able to implement adjustment strategies more quickly than public-sector firms. In the private sector the incentives to pursue adjustment strategies were related to profitability and the survival of firms, whereas in the public sector the incentives to adjust were part of a broader process of political and institutional adjustment. The private sector, with its profit maximization ethos, was more adversely affected by the introduction of more competitive procurement policies (as opposed to the former cost-plus policy) than the public sector, and therefore pursued adjustment strategies more vigorously than public-sector defence firms.

The impact of the defence cuts was uneven across the private-sector arms industry because of differences in company size, products, degree of dependence on defence sales and position on the product hierarchy. In response to the defence cuts most defence firms were prompted to pursue various adjustment strategies, defensive, offensive or both. The most common offensive adjustment strategies included the concentration and monopolization of certain defence markets, increasing exports, and joint ventures and alliances with foreign defence firms. These offensive strategies had a number of significant outcomes. The domestic defence market became characterized by higher levels of concentration, as certain large firms (such as Reumech) were able to consolidate or extend their monopoly positions in certain sectors of the domestic defence market. Some firms (such as XCEL) were able to increase the value and share of their defence business through shifting demand patterns in the domestic defence market (i.e., greater expenditure on maintenance rather than new procurement).

Most private-sector defence firms embarked on export drives to offset the declining domestic demand for armaments and to find new markets for their defence products. Although there were some barriers to exporting, such as UN Security Council Resolution 558 of 1984, which encouraged countries to refrain from purchasing South African armaments, many firms were successful

---

[74] *Some Statistics of the SADIA Members of the Defence Industry* (note 65).

[75] The low value of private-sector arms exports is also a function of the fact that Denel, as the major defence contractor, is the primary seller of defence systems, and therefore the value of exports is reflected more directly in Denel's financial results. Most private-sector firms would act as subcontractors and suppliers for Denel's export products.

in finding new export markets for their defence products. Larger firms, such as Plessey and Reutech, which had extensive links with multinational companies, were relatively successful in finding export markets for their products. Smaller firms with niche products, such as Milkor, also managed to find export markets, while in some firms arms exports exceeded the value of local defence business or replaced local defence business (as in the case of Northbend). Most firms also pursued joint ventures and strategic alliances with foreign defence firms.

The most common defensive adjustment strategies involved laying off workers, down-sizing defence activities and diversification. The private-sector arms industry down-sized quite dramatically after 1989 and this was reflected in the loss of over 45 000 jobs by 1993. Cut-backs on R&D and capital expenditure as part of down-sizing contributed to the underutilization of resources (capital, labour and technology) in the manufacturing sector: many defence firms were operating at suboptimal production levels with large surplus capacities. Many firms adopted spin-off or diversification strategies. Very few attempted to convert to civilian production. In terms of spin-off strategies, some larger firms (e.g., Grinaker Avitronics or Reumech) were relatively successful in developing civilian products from their existing military technologies, whereas smaller firms found it more difficult to develop spin-offs because of the resources needed to fund R&D to develop new civilian products. In terms of diversification, many larger firms such as Reumech were successful in acquiring civilian firms or civilian product lines through licensing agreements.

In terms of general trends, most of the medium-sized and large firms in the private sector which had a high dependence on defence sales in the late 1980s managed to weather the transition to lower levels of demand through various adjustment strategies such as diversification. These firms saw real increases in turnover and civilian sales after 1989. However, many small and medium-sized firms in the private sector which had a high dependence on defence sales in the late 1980s found the adjustment more painful. Many of these experienced real declines in turnover and defence sales after 1989 and by April 1994 were struggling to survive.

A number of factors at the level of individual defence firms and certain macroeconomic-level factors inhibited the success of many private defence firms' adjustment efforts. The specific technological, organizational and cultural features of individual defence firms meant that exit from the defence market was often difficult and costly.[76] The barriers to exit included the highly protected nature of the defence market, inexperience in commercial marketing, an emphasis on product rather than process innovation and a management culture isolated from commercial practices. Macroeconomic factors, such as the severe domestic recession, together with the absence of public policies to support defence firms' adjustment efforts or to encourage the further diversifica-

[76] DiFilippo, A., *How the Military Serving Firm Differs from the Rest*, ECD Briefing Paper no. 10 (National Commission for Economic Conversion and Disarmament: Washington, DC, 1991) highlights the differences between defence and civilian firms. Gansler, J., *Affording Defence* (MIT Press: Cambridge, Mass., 1989) highlights the barriers to entry to and exit from the defence market.

tion or conversion of South Africa's arms industry, also inhibited these adjustment efforts. The uncertainty surrounding South Africa's transition to majority rule and the government's lack of a clearly defined defence industrial policy made the difficulty worse.

## IV. Conclusions

This chapter examines the adjustment experience of South Africa's arms industry during the process of disarmament between 1989 and April 1994. As a result of the dramatic cuts in defence expenditure, and particularly procurement expenditure, between 1989 and 1994, the industry underwent down-sizing and restructuring. In response to the decline in the domestic defence market, and in order to survive, most defence firms, both public- and private-sector, were forced to pursue various adjustment strategies. The most popular included increasing exports, concentration and monopolization of defence markets, and diversification.

The wide range of adjustment strategies pursued by defence firms reflected the absence of any coherent government strategy towards the down-sizing of the arms industry. It also reflected the fact that defence firms were forced to rely on their own resources and initiatives to survive the impact of the defence drawdown, thereby making fragmentation of the arms industry likely. As a result of these adjustment strategies, which included large-scale job losses, many public- and private-sector defence firms were able to survive the effects of the shrinking domestic defence market. However, a number of private-sector defence firms exited the defence market between 1989 and 1994, either by selling or by closing down their defence operations.

When the ANC-led government came to power in April 1994, it inherited a domestic arms industry which was substantially smaller and more concentrated but less cohesive than it had been during the late 1980s. Despite the breakdown of the cohesiveness of the apartheid-era military–industrial complex, it managed to retain a certain amount of residual political influence. This was reflected in the arms industry's (via Armscor) dominant role in the negotiations between the ANC and the government on defence matters in 1993 and 1994 and in its continued access to executive levels of the state.

Despite the implementation of substantial defence cuts and disarmament measures, at no point during the period between 1989 and April 1994 did the de Klerk Government attempt to support the adjustment efforts of defence firms with either macroeconomic or micro-economic initiatives, such as subsidies for retraining of defence workers, or to implement a national conversion and industrial strategy. The absence of any government initiatives for defence industrial adjustment meant that valuable skills and technologies which the arms industry embodied were either lost or wasted as firms attempted to down-size and adjust to the shrinking domestic defence market.

# 6. The changing security environment after 1994[*]

## I. Introduction

The elections of 27 April 1994 mark a watershed in South Africa's violent history. The end of white minority rule and the assertion of democracy have brought with them a transformation of the machinery of the South African state. Following decades of unaccountability and lack of transparency, policy procedures have been opened up to widespread public scrutiny and democratic participation. These measures have triggered a profound normative, institutional and cultural transformation of all sectors of South African society.

The military institutions of the apartheid state have been a specific target for reform because of the central role they played in the apartheid system and because of the strong undercurrent of anti-militarism which infuses the ANC, particularly at the grass-roots level. There is a loose grouping of politicians, academics, lawyers, policy makers and church members whose voices have gained prominence in the national security debate as a result of the democratic transformation of South African society. They are vociferous advocates of demilitarization and in many ways have had an inordinate influence on the defence and security policy debates which have occurred since the transfer of power to the Government of National Unity.[1] They argue for a dramatic change in South Africa's defence and security policies by broadening the concept of security to encompass non-military factors such as economic, environmental and human security.[2] Here the stress lies on developmental and collective approaches to regional security challenges.[3] Respect for human rights and observance of international norms lie at the heart of this approach. Kader Asmal, a cabinet member in the new government, has argued that 'without dis-

---

[1] This group includes members of the ANC's Military Research Group, Black Sash (lawyers), numerous ANC politicians, notables such as Dr Ian Philips, former member of the Standing Committee on Defence, the Anglican Church headed by Bishop Desmond Tutu, non-governmental organizations (NGOs) such as the Group on Environmental Monitoring, academics such as Maxi van Aardt, Gavin Cawthra, Jacky Cock, Laurie Nathan, Peter Vale and and many others too numerous to name.

[2] For a discussion of these issues, see van Aardt, M., 'In search of a more adequate concept of security for South Africa', *South African Journal of International Affairs*, vol. 1, no. 1 (spring 1993), pp. 82–101.

[3] Booth, K. and Vale, P., 'Security in Southern Africa: after apartheid, beyond realism', *International Affairs*, vol. 71, no. 2 (1995), pp. 285–304.

[*] In this chapter the work of Jakkie Cilliers, Greg Mills, Laurie Nathan, Pierre Steyn, Peter Vale and Rocky Williams is extensively referred to. Not only are they South Africa's foremost security and foreign relations analysts, having published widely on these subjects; more significantly they play an important role in policy formulation, collectively exerting considerable influence on the policy debate. Both the authors have held numerous discussions with this group of individuals over the past four years, which has provided much insight into the subjects dealt with in this chapter.

counting the element of national interest as one of the motivating factors for a state's behaviour, a free South Africa must display a deep attachment to international norms of behaviour based on law. In this way, we can repudiate the behaviour of the past and insert a new morality where hegemony, domination and over-lordship give way to good neighbourliness, mutual respect and solidarity'.[4] This preoccupation with the moral and ethical dimensions of security extends to the debate on the future of the domestic arms industry and the arms trade.[5]

As a consequence of these influences on the policy debate the security establishment has found itself in an equivocal position. It faces the difficult task of realigning itself within the context of policy changes imposed by a newly empowered civil authority which does not always hold its interests close at heart. No part of the South African defence establishment has remained untouched by the fundamental upheavals associated with South Africa's transition to democracy. Since 1994, greater transparency and accountability in military affairs have been imposed through the creation of a civilian-run Ministry of Defence and a powerful parliamentary Defence Select Committee which has powers to define policy.[6] Moreover, the integration of the previously hostile 'statutory' and 'non-statutory' military forces into a newly created South African National Defence Force (SANDF) in 1994 has produced a new military entity more representative of South African society as a whole. At the same time security policy, strategic doctrine and operational practice have all been subject to a process of re-evaluation.[7] The first stage involved the drafting of the 1996 White Paper on National Defence which laid out the broad conceptual framework for defence and security policy.[8] The second stage, the Defence Review, began in February 1996, the review being approved by the cabinet in June 1997 and by parliament in August 1997.[9] It attempted to translate the conceptual framework of the White Paper into the nuts and bolts of institutional and operational practice. The final stage, which started in February 1997, aims to produce a White Paper on the Defence Industry which is consistent with the

[4] Asmal, K., 'The role of the armed forces in a new South Africa in the context of international law', Paper presented at the Conference on Civil–Military Relations in a Post-Settlement South Africa, Institute for Defence Policy, Pretoria, Apr. 1992.

[5] For international human rights guidelines on South Africa's arms controls, see South African Department of Defence, *Defence in a Democracy: White Paper on National Defence* (Government Printer: Pretoria, 1996), pp. 41–45.

[6] For further details, see chapter 7 in this volume.

[7] For an overview of these issues in the process of transformation, see Nathan, L., *Changing the Guard: Armed Forces and Defence Policy in a Democratic South Africa* (Human Sciences Research Council: Pretoria, 1995).

[8] *White Paper on National Defence* (note 5). See also chapter 1 in this volume.

[9] 'SA review approval may be too late', *Jane's Defence Weekly,* 2 July 1997, p. 15; and 'Guns, butter, peace and security', *Sunday Times* (Johannesburg), 14 Sep. 1997. So far no formal document has been produced on the Defence Review, but two draft documents have been issued: Defence Secretariat, Draft Defence Review Document, 1 Aug. 1996; and Defence Secretariat, Draft Defence Review Document, 15 Feb. 1997.

1996 White Paper and the Defence Review.[10] The outcome of this process will have long-term implications for the South African defence industrial base.

This process of transformation has contributed to an atmosphere of uncertainty for the country's military.[11] Colonel Rocky Williams, a strategic analyst within the new Defence Secretariat in Pretoria, has observed of this process that 'a period of profound crisis faces the South African military. Domestically this relates to the exigencies of the current transition and the manner in which this has impacted on the institutional existence of the armed forces. Internationally and regionally it derives from the end of the Cold War and the diminution of armed forces' profile within the Southern African region'.[12]

The radical changes proposed by the adherents of demilitarization have not gone unchallenged either by the military or by the domestic arms industry. Representatives from Armscor, the former SADF, the domestic arms industry and hawkish and influential elements within the ANC, such as Joe Modise, the new Minister of Defence, have strongly emphasized that the backbone of a strong and independent state is a robust military supported by a vibrant arms industrial base. Even so, despite the apparent polarization of views between the 'doves' and the 'hawks', the spirit of reconciliation that has come to characterize South Africa's remarkable transition has resulted in a degree of compromise in security matters on both sides of the debate.

Rapprochement may have modified the extent of demilitarization within South African society but the fact that the military and the arms industry have been forced to react to an agenda which has largely been defined by non-military actors signifies the difference between the post-apartheid era and the previous period of disarmament under the de Klerk Administration. Moreover, the gathering pace of change at the regional level and the re-entry of South Africa into the international community have forced a major examination of South Africa's priorities in the foreign and security policy field which are now strongly influenced by non-traditional security challenges.

This chapter demonstrates the profound implications which the realignment of South Africa's foreign and security policies since the ending of apartheid has for the future role of the military and South Africa's defence industrial base. In capturing these changes it examines the recasting of the security agenda, emerging forms of regional security cooperation, peacekeeping, new arms sales policies, strategic doctrine, the military backlash and the implications for the arms industry.

[10] Throughout the process of reforming South Africa's security establishment, one feature unique to the South African experience requires a special mention, namely the degree of open consultation on defence and security policy. No group or individual was barred from submitting both written and oral presentations on aspects of national defence and security policy. Moreover, all contributions were treated with equal consideration and respect by the teams appointed to write up the policy documents. The military, used to controlling their own destiny, found themselves engaged with civil society in a highly charged but nevertheless healthy public debate about all aspects of defence and security policy.

[11] Williams, R., 'Non-offensive defence and South Africa: considerations on a post-modern military, mission redefinition and defensive restructuring', *NoD and Conversion*, no. 37 (July 1996). This article was first produced as an internal discussion document within the Defence Secretariat in Pretoria.

[12] Williams (note 11) p. 28.

## II. Recasting the security agenda

With the demise of apartheid and the withdrawal of the superpowers from the Southern African region, South Africa has been attempting to redefine its external relations both within the region and with the rest of the world. South Africa has witnessed a dramatic rethinking of its defence and security policies.[13] This rethinking has not been confined to theoretical debates but has had a very real impact on the formulation and implementation of new security policies in the post-apartheid era. It was facilitated by the general transformation which began in the region, both at a national level through democratization and the ending of civil wars and at the regional level through cooperation between states since the ending of South Africa's policies of military aggression and regional destabilization.[14]

During the apartheid era, security figured as an integral part of the white minority's discourse in which their country was cast as an embattled bastion of Western civilization in the global struggle against communism and the anarchy of black Africa.[15] The powerful symbolism of civilization under attack helped propel the military to the centre stage of policy making within the apartheid state. Booth and Vale have argued, however, that the centrality of militarism to South Africa's security policy helped to legitimize a ruthless regime which consistently undermined the security of its neighbours and its own citizens.[16] The 'destabilization' campaign of the 1980s resulted in the deaths of more than 1 million people and is estimated to have cost regional GDP some $62.42 billion.[17] Ultimately, South Africa's security policy proved dysfunctional to the apartheid state in that it created such internal instability that the security of the white minority was threatened.

The reformulation of the concept of security has emerged in response to the challenge to the established principles and practice that dominated the policies of the Ministry of Defence and Department of Foreign Affairs during the apartheid era. For these institutions, during the apartheid era the notion of security was subsumed under the rubric of power and was synonymous with the security of the state against external threat, which was perceived as being achieved through enhancing military capabilities. The emphasis on a state-centred definition of security grew out of the realist assumptions, which underpinned apartheid security thinking, of a sharp boundary between domestic order and international anarchy (the Hobbesian state of nature) where war is an ever-present possibility.[18] The utility of realist notions for maintaining security in the developing world has, however, been highly contested. 'From a Southern per-

[13] van Aardt, M., 'Doing battle with security: a Southern African approach', *South African Journal of International Affairs*, vol. 3, no. 2 (summer 1996), p. 21.
[14] See chapter 3 in this volume.
[15] Booth and Vale (note 3), p. 287.
[16] Booth and Vale (note 3), p. 287.
[17] Booth and Vale (note 3), p. 286.
[18] For a classical explanation of realism, see Waltz, K., *Theory of International Relations* (Addison-Wesley: Reading, Mass., 1979), p. 102.

spective, military conflicts are rarely cross-border, but rather, the result of domestic challenges to the legitimacy of political regimes frequently supported by outside intervention.'[19] As in the case of apartheid South Africa, for many citizens the strong military state, far from providing security, represented the single largest threat to their personal security. South African security forces were regularly used in internal security operations within the townships and squatter camps during the apartheid years.

The profound changes which have taken place in the global and regional strategic environment since 1989 have put the parameters of South Africa's realist doctrine to the test. Not only has the notion of an external military threat disappeared with the demise of communism; the notion of being a beleaguered outpost of Western civilization has evaporated with the current political and cultural transformation.[20] By being accepted back into the international community South Africa is also being exposed more directly to global trends and influences such as economic interdependence (through trade, investment and financial services, global telecommunications and media networks), which are all contributing to an undermining of the notions of sovereignty and national interests that formed the basis of South Africa's traditional approach to security policy and practice.[21]

The new government has articulated the view that South Africa's greatest security challenge is its development crisis.[22] 'The real threats to [South Africa's] security are internal and non-military—poverty, underdevelopment, unemployment, illiteracy, inadequate access to basic services like water and health, and rising levels of crime and violence.'[23] For most people in South Africa the feeling of insecurity has always arisen from fears about hunger, disease and deprivation rather than from threats to national sovereignty. Poverty affects the lives of millions of South Africans. It is estimated that 17 million people live below the minimum subsistence level.[24] Income distribution remains one of the most unequal in the world: the per capita income of whites is around 12 times higher that of blacks, the incomes of the richest households being some 45 times higher than those of the poorest 20 per cent. Unemployment remains high, particularly in the black community, where it is estimated to be 50 per cent.[25] It is forecast that at current rates of growth (the population is anticipated to rise to 58 million in 2010 and 73 million in 2025) jobs will be

[19] For a critique of the realist point of view from a Southern perspective, see Tickner, J., 'Re-visioning security', eds K. Booth and S. Smith, *International Relations Theory Today* (Polity Press: London, 1995), p. 179.

[20] Booth and Vale (note 3), p. 287.

[21] For a general discussion of the limits of the realist paradigm in the context of the post-cold war era, see Halliday, F., 'The end of the cold war and international relations: some analytical and theoretical conclusions', eds Booth and Smith (note 19), pp. 38–61.

[22] *White Paper on National Defence* (note 5), p. 5. See also Booth and Vale (note 3).

[23] 'Trimming back', *Financial Mail*, 21 Mar. 1997.

[24] African National Congress, *The Reconstruction and Development Programme: A Policy Framework* (Umanyano Publications: Johannesburg, 1994), p. 14.

[25] All figures in this section are taken from Handley, A. and Mills, G. (eds), *From Isolation to Integration: The South African Economy in the 1990s* (South African Institute of International Affairs: Johannesburg, 1996), p. 2.

found for only 7 per cent of newcomers to the job market, compared to an absorption rate of 80–85 per cent in 1965–70. Approximately half of the current unemployed are under the age of 30 and almost 90 per cent lack skills or training of any kind. Some 12.5 million or 30 per cent of the population are believed to be illiterate. These conditions provide fertile ground for future unrest if they are ignored. The government has stated that 'the RDP is the principal long term means of promoting the well being and security of citizens and, thereby, the long term stability and development of the country. There is consequently a compelling need to reallocate state resources to the RDP'.[26]

Economic constraints and pressing development needs combined with the high expectations of the black majority do not augur well for long-term internal stability. Even if South Africa maintains its present rate of growth at just above 3 per cent in 1995, with a current population growth of 2.5 per cent and unemployment at nearly 50 per cent of the labour force the potential for redistributing wealth to overcome the inequalities of the past is limited.[27] South Africa's future economic performance will have a profound effect on national and regional stability.

The security problems that South Africa currently faces are also compounded by the spillover effects of the security problems of neighbouring states. The demise of the apartheid regime, although eradicating the major cause of regional conflict, has not been accompanied by a decline in the overwhelming range of security problems in Southern Africa for which no military solution is apparent.

Most people in Southern Africa continue to experience extremes of insecurity as a result of the legacies of apartheid.[28] In war-torn Angola and Mozambique the widespread displacement of peoples and the subsequent disruption of economic activities, in particular agriculture, have left tens of thousands dependent on food aid for survival. Large numbers of refugees, particularly from Mozambique, have found their way into South Africa, intensifying the problems of urban poverty and unemployment and contributing to growing xenophobia and ethnic tensions within the townships. The widespread dispersal of landmines means that much agricultural land in Angola and Mozambique will remain uninhabitable and uncultivable for the foreseeable future, delaying the return of refugees and internally displaced peoples.

The severe drought of 1992 intensified the situation of food scarcity and exacerbated regional migration patterns. Arguably, food insecurity is a more fundamental threat to the region than the likelihood of interstate violence, but the achievement of food security requires a far higher level of regional cooperation than exists at present. A crisis in the supply of one or more basic necessities (such as water) may precipitate further disasters. For example, when food scarcity and water shortages coincided with the widespread displacement of

[26] *White Paper on National Defence* (note 5), p. 5.
[27] *South Africa: Country Survey, 2nd Quarter* (Economist Intelligence Unit: London, 1996).
[28] Mills, G., 'South Africa and Africa: regional integration and security co-operation', Paper presented at the AIC Conference on Defence, Midrand, 11–12 Oct. 1994.

populations and the breakdown of communities in Mozambique during the latter part of the war epidemics of fatal diseases such as cholera, typhoid and meningitis broke out, with devastating effects on the young, the old and the weak.[29]

The presence of a huge amount of war *matériel* resulting from civil wars and South Africa's wars with the Front-Line States also provides a continued source of instability within the region. Small-arms flows into and within the region, and particularly into South Africa from Angola, Mozambique and Namibia, are facilitating an unprecedented wave of violent crime in the townships and city centres, intensifying the sense of individual human insecurity.[30] In war-torn countries such as Angola and Mozambique the widespread presence of small arms potentially threatens the fragile process of post-conflict peace-building.[31] As Christopher Clapham has commented, 'war teaches few skills beyond the use of weapons: it destroys much of the already weak economic base on which a newly independent government must painfully build; and fighters who view themselves as having borne the brunt of the struggle for freedom, then find their expectations of victory bitterly disappointed, have few resources with which to improve themselves beyond a renewed resort to arms'.[32]

Whether South Africa likes it or not it cannot escape the problems of its neighbours because they directly impinge upon its own stability and long-term security. Traditional military approaches to security are powerless to confront these contemporary challenges to regional stability.[33] 'It is realistic to suggest that a broad-based development orientated approach to building regional security would address these kinds of socio-economic issues better than a purely militaristic response.'[34] The development approach to security finds expression in the UNDP *Human Development Report 1994*, which argued that:

The world can never be at peace unless people have security in their daily lives. Future conflicts may often be within nations rather than between them with their origins buried deep in growing socio-economic deprivation and disparities. The search for security in such a milieu lies in development, not in arms. More generally it is not possible for the community of nations to achieve any of its major goals—not peace, not environmental protection, not human rights or democratisation, not fertility reduction,

---

[29] Willett, S., 'Ostriches, wise old elephants and economic reconstruction in Mozambique', *International Peacekeeping*, vol. 2, no. 1 (spring 1995), p. 40.

[30] Cock, J., 'Light weapons proliferation in Southern Africa as a social issue', Paper presented at the Arms Industry Workshop, Group for Environmental Monitoring and Centre for Conflict Resolution, Johannesburg, 14–15 Feb. 1996.

[31] Batchelor, P., Potgieter, J. and Smith, C., *Small Arms Management and Peacekeeping in Southern Africa* (United Nations Institute for Disarmament Research: Geneva, 1996).

[32] Clapham, C., 'The African state', Paper presented at the Conference of the Royal African Society on Sub-Saharan Africa: The Record and Outlook, Cambridge, Apr. 1991.

[33] For a discussion of the regional security challenges and the limits of military intervention, see Cilliers, J., 'Towards collaborative and co-operative security in Southern Africa: the OAU and SADC', eds J. Cilliers and M. Reichardt, *About Turn: The Transformation of the South African Military and Intelligence* (Institute for Defence Policy: Halfway House, 1996).

[34] Mills (note 28), p. 75.

not social integration—except in the context of sustainable development that leads to human security.[35]

The development approach to regional stability has been persistently championed by a number of influential ANC foreign- and security-policy advisers.[36] In 1996 these alternative approaches to security were given official recognition in the White Paper on National Defence, which defines South Africa's new approach to security as 'an all encompassing condition in which individual citizens live in freedom, peace and safety, participate fully in the process of governance, enjoy the protection of fundamental rights, have access to resources and the basic necessities of life, and inhabit an environment which is not detrimental to their health and well being'.[37] There is also stress on aspects of human security, reflecting the ANC's strong preoccupation with human rights. 'National security shall be sought primarily through efforts to meet the political, economic, social and cultural rights and needs of South Africa's people.'[38]

The problem with the White Paper's broader concept of security is that it is not always clear how it will be implemented in practical or operational terms, especially in relation to societal, economic and environmental security.[39] Nowhere is this more apparent than at the regional level where attempts to cooperate on this broadly based notion of security have, so far, floundered as a result of lack of resources, expertise and operational procedures.

## III. Regional security cooperation

Regional cooperation on security issues also takes a prominent place in the White Paper. It emphasizes the development of a common security approach through the SADC and acknowledges that many of the states in the region share the same domestic threats with potential for spillover into neighbouring countries. By placing emphasis on collective forms of security the new security thinking has highlighted the importance of economic, political and social forms of cooperation, while focusing security policy on preventive diplomacy, socio-economic development and conventional arms limitations.[40]

Attempts at regional security cooperation have centred on the SADC. Its founding document provided the following motivations for its mandate on political, military and security issues: 'War and insecurity are the enemy of economic progress and social welfare. Good and strengthened political relations among countries of the region, and peace and mutual security, are critical components of the total environment for regional cooperation and integration. The

---

[35] United Nations Development Programme, *Human Development Report 1994* (Oxford University Press: Oxford, 1994), p. 1.

[36] Booth and Vale (note 3); and Nathan, L. and Honwana, J., *After the Storm: Common Security and Conflict Resolution in Southern Africa*, Arusha Papers, no. 3 (University of the Western Cape, Centre for Southern African Studies: Cape Town, 1995).

[37] *White Paper on National Defence* (note 5), p. 5.

[38] *White Paper on National Defence* (note 5), p. 6.

[39] van Aardt (note 13), p. 21.

[40] Nathan and Honwana (note 36).

region needs, therefore, to establish a framework and mechanisms to strengthen regional solidarity and provide for mutual peace and security'.[41]

A 1993 SADC document, 'A Framework and Strategy for Building the Community', stated that 'a framework must be found for enhancing political solidarity and harmony among member states in order to minimize conditions which can lead to national and regional instability and insecurity'.[42] A number of strategies for advancing peace and security in the region were listed, such as the creation of a forum for mediation and arbitration, reductions in force levels and military expenditure, the adoption of non-offensive defence doctrines, and the implementation of confidence- and security-building measures.[43]

The formalization of the SADC's security aspirations took place in January 1996, when SADC ministers for defence and foreign affairs met in Gaborone, Botswana, to agree the terms of reference for the setting up of a SADC Organ on Politics, Defence and Security.[44] The ministers recommended that 'in view of the reduced tensions in the region, military force levels and expenditure should be reduced to the minimum level required for territorial defence'.[45]

The Inter-State Defence and Security Committee (ISDSC), established in 1983 under the aegis of the Front-Line States organization, has been rejuvenated and now forms part of the Organ on Politics, Defence and Security.[46] The ISDSC, which operates as an informal structure and is chaired by SADC defence ministers on a rotational basis, is made up of a ministerial council, three subcommittees (for defence, security and intelligence) and a number of sub-subcommittees dealing with operational, intelligence, aviation, medical and other matters.[47] The task of these committees is to examine, advocate and coordinate activities on an extensive range of issues including cross-border crime, illegal immigrants, peace support operations and intelligence sharing. They convene on a frequent basis, but have so far only operated in a somewhat ad hoc manner.

The progress made in military forms of cooperation may be explained partly by the fact that the military has at its disposal far more resources than other parts of the SADC machinery. The role of the SANDF is particularly important in this respect. It has been quick to focus on military forms of regional coop-

[41] Southern African Development Community, Towards the Southern African Development Community: A Declaration by the Heads of State or Government of Southern African States, Windhoek, 1992, pp. 9–10.
[42] Southern African Development Community, Southern Africa: A Framework and Strategy for Building the Community, SADC, Harare, Jan. 1993.
[43] There are many problems with the SADC approach to regional security, not least the question who the primary referents of regional security are—for instance, the state or people? This issue is discussed in some depth by van Aardt (note 13), pp. 23–26.
[44] Southern African Development Community, The SADC Organ on Politics, Defence and Security. Statement issued by the Meeting of SADC Ministers Responsible for Foreign Affairs, Defence and SADC Affairs, Gaborone, 18 Jan. 1996.
[45] Quoted in Williams (note 11), p. 45.
[46] Cilliers, J., 'The evolving security architecture in Southern Africa', African Security Review, vol. 4, no. 5 (1995), p. 41.
[47] Cilliers, J. and Malan, M., SADC Organ on Politics, Defence and Security: Future Development, ISS Papers no. 19 (Institute of Strategic Studies: Halfway House, Mar. 1997), p. 2. The Institute of Strategic Studies was formerly the Institute for Defence Policy, renamed in early 1997.

eration, such as peacekeeping, military assistance and training programmes, and disaster relief, and by offering the experience of Armscor as a regional arms procurement agency. Exchanges of personnel and observers, combined exercises and the opening up of its training establishments are some of the practical measures which the SANDF has promoted at the regional level.

The narrow focus of the military's preoccupations results in an agenda which appears much more manageable than one which promotes an expanded notion of security. This has the effect of making the military's arguments appear more concise and its goals more attainable than the arguably more pressing concerns of human and environmental security within the region. Progress towards achieving the broader goals of collective security is meanwhile constrained by the lack of established procedures and mechanisms and the absence of the skills and techniques required for conflict resolution and mediation.

Ultimately the potential for realizing the SADC's broader security goals depends upon establishing a genuine collective identity at the regional level. This in itself is fraught with pitfalls. During the apartheid era a common identity of purpose was forged between the Front-Line States (the originators of the SADC) because of the existence of a common enemy, namely apartheid South Africa. With the demise of apartheid, it is questionable whether there is a binding force that can any longer sustain a collective sense of purpose beyond a loose alliance of nation-states bound in a common geographical area.

With South Africa now a key actor in the evolution of any future regional security arrangement, the convergence and complementarity of South Africa's interests with those of its neighbours are central to the issue of regional integration and a collective sense of purpose.[48] There is a sense in which the other states in Southern Africa believe they can be South Africa's equals within the SADC, when in fact the reality of South Africa's overwhelming economic and military power exposes this illusion.[49] Even with the most cooperative will in the world South Africa will inevitably set the regional agenda, and it will primarily reflect South Africa's interests. Thus, although the security debate highlights regional integration, there is growing concern on the part of South Africa's neighbours that the evolution of the SADC may weaken national sovereignty and undermine their national interests. Residual fears about South Africa's hegemonic intentions and rivalry for regional influence all contribute to an underlying unease within the SADC framework. In particular, President Robert Mugabe of Zimbabwe, who was the leader of the Front-Line States, is believed to be resentful at his position as a regional leader being usurped by the dominance of South Africa in regional security issues.[50]

[48] For a discussion on the limits to integration, see Solomon, H. and Cilliers, J. (eds), *People, Poverty and Peace: Human Security in Southern Africa*, IDP Monograph Series no. 4 (Institute for Defence Policy: Halfway House, May 1996).

[49] George, P., 'The impact of South Africa's arms sales policy on regional military expenditure, development and security', ed. Swedish Ministry of Foreign Affairs, *Säkerhet och utveckling i Afrika* [Security and development in Africa] (Utrikesdepartementet: Stockholm, Nov. 1995), appendix, p. 255 [appendix in English].

[50] Cilliers (note 46), p. 41.

**Table 6.1.** Military expenditure and armed forces in Southern Africa, 1996

|  | Military expenditure (US $m., constant 1995 prices) | Armed forces | MBTs | APCs | AFVs | Art | Cbt air | Ships |
|---|---|---|---|---|---|---|---|---|
| Angola | 300 | 97 000 | 400 | 100 | 150 | 300 | 36 | 6 |
| Botswana | 225 | 7 500 | 36 | 30 | – | 16 | 34 | .. |
| Lesotho | 33 | 2 000 | – | – | – | 2 | – | .. |
| Malawi | 21 | 9 800 | – | – | – | 9 | – | .. |
| Mozambique | 58 | 11 000 | 80 | 250 | 40 | 100 | 43 | 3 |
| Namibia | 65 | 8 100 | – | 20 | – | – | – | – |
| South Africa | 3 720 | 137 900 | 250 | 1 660 | 1 500 | 350 | 234 | 12 |
| Tanzania | 87 | 34 600 | 95 | 100 | – | 300 | 24 | 16 |
| Zambia | 62 | 21 600 | 30 | 13 | – | 100 | 59 | .. |
| Zimbabwe | 233 | 43 000 | 40 | 105 | – | 16 | 58 | .. |

*Notes:* MBT = main battle tank; APC = armoured personnel carrier; AFV = armoured fighting vehicle; Art = artillery; Cbt air = combat aircraft. Ships include coastal and patrol ships and combatants.

*Source:* International Institute for Strategic Studies, *The Military Balance 1996–1997* (Oxford University Press: Oxford, 1997).

The road to closer regional cooperation on security issues will be long and hard, as its success depends on the complex interaction of a number of variables. Genuine regional cooperation on security matters can only take place once South Africa's neighbours overcome their mistrust of its regional intentions. South Africa's present overwhelming military balance of power continues to feed distrust in the region, which suggests that it has some way to go towards building a sense of confidence and security with its neighbours (see table 6.1).

South Africa is the largest military spender in absolute terms in Southern Africa, and its military spending accounts for nearly 80 per cent of the region's total. It currently has the largest and most sophisticated armed force in Southern Africa and the SANDF's inventory of military equipment is significantly better in qualitative, quantitative and operational terms than those of all the other countries in the region put together. These factors taken together feed into the residual fears about South Africa's hegemonic intentions within the region and place the onus on South Africa to take the initiative in building trust and confidence about its future regional ambitions.

Partly as a consequence of these fears, common security efforts are confined to ad hoc cooperative measures rather than the establishment of a formal security alliance. Even in those areas in which it looks likely that there will be deeper levels of interaction in the future, for example in conflict resolution and peacekeeping, policies are evolving in a haphazard manner with little support among members for a formalized decision-making and institutional framework.

## IV. Peacekeeping

Peacekeeping is a major means by which the South African military has been promoting its future role in both the regional and the continental context. This aspiration has coincided with considerable external pressure on South Africa to adopt a continental peacekeeping role, particularly as a result of the escalation of conflicts in Zaire and the Great Lakes region. Following initial enthusiasm for this role, the SANDF has become more circumspect about South Africa's participation in peace support operations, particularly while it is undergoing the politically sensitive process of force integration. It soon became apparent during initial training with the British Military Assistance Training Team (BMATT) that the SANDF lacked experience of, training for or skill in the requirements of the complex tasks of peacekeeping.[51] There is growing awareness of the liabilities of peacekeeping[52] and increasing recognition that international pressure could lead South Africa into protracted (and costly) conflict situations from which it might be difficult to withdraw.

The 1996 White Paper recognizes that as a fully fledged member of the international community South Africa has an obligation to participate in international peace support operations. However, it is clear that its participation will only be considered under certain strict conditions. These include:

1. There should be parliamentary approval and public support for such involvement. This will require an appreciation of the associated costs and risks, including financial costs and risks to the lives of military personnel.

2. The operation should have a clear mandate, mission and objectives.

3. There should be realistic criteria for terminating the operation.

4. The operation should be authorized by the UN Security Council.

5. Operations in Southern Africa should be sanctioned by the SADC and undertaken with other SADC states rather than be conducted on a unilateral basis. Similarly, operations in Africa should be sanctioned by the Organization of African Unity (OAU).[53]

Despite this caution the first tentative steps have been taken towards establishing a regional peacekeeping force. In April 1997 soldiers from nine Southern African countries, including South Africa, took part in peacekeeping training in Nyanga in north-eastern Zimbabwe under the expert guidance of BMATT. South Africa's participation is highly significant given the role of its troops within the region less than a decade ago. Funding for the programme was provided by the British Foreign and Commonwealth Office, which contributed £300 000, and the Zimbabwean Government, which matched this sum.[54] Major

---

[51] Interview with the Second Political Secretary, British High Commission, Pretoria, 12 Nov. 1996.

[52] Shaw, M. and Cilliers, J. (eds), *South Africa and Peacekeeping in Africa, Vol. 1* (Institute for Defence Policy: Halfway House, 1995), pp. 9–10.

[53] *White Paper on National Defence* (note 5), p. 30.

[54] Meldrum, A., 'Africans practise art of regional peacekeeping', *The Guardian*, 14 Apr. 1996.

Cobus Valentine of the SANDF is reported to have said of this process that 'this is the first time we are part of a regional peace keeping force . . . we are building a foundation of understanding and cooperation for the future . . . this will only get stronger'.[55] Clearly, the initial stages of establishing a regional peace-keeping force have acted as a powerful confidence- and security-building mechanism by bringing former adversaries together to work towards the common cause of peace. However, there is a long way to go before an integrated regional peacekeeping force will be ready for action on the African continent.

In addition to involvement in regional peacekeeping initiatives organized under the auspices of the SADC, the South African Government has become involved with initiatives instigated by the OAU. Presenting South Africa's foreign policy priorities to the portfolio Committee on Foreign Affairs in parliament in March 1995, Foreign Minister Alfred Nzo declared that 'any South African involvement in the prevention or solving of conflict solutions elsewhere in Africa, should take place within the framework of the OAU's Mechanism for Conflict Prevention, Management and Resolution. Only if the OAU is seen to be accepting responsibility for, and dealing with its own problems, will the Organisation and our Continent earn the respect of outsiders'.[56] The Central Mechanism for Conflict Prevention, Management and Resolution was established by the OAU on 30 June 1993. The Declaration of the Assembly of Heads of State and Government on the establishment of the mechanism committed the OAU to close cooperation with the UN in respect of peacemaking and peace-keeping. At the same time the mechanism is committed to cooperation with regional organizations such as the SADC. So far, however, it has not been a notable success in conflict prevention or management on the African continent. Election observation has been its most active area, in addition to mediation in a number of conflict situations in sub-Saharan Africa.

So far the OAU has placed greatest emphasis on preventive diplomacy, deciding that peacekeeping should not constitute a major activity of the organization. This decision is largely influenced by the OAU's lack of resources. Conflict resolution and peacemaking through diplomacy are far more cost-effective areas of engagement for an organization seriously short of money. Despite its obvious lack of resources, the OAU remains under considerable international pressure to play a more proactive role in African peacekeeping.

## V. Arms sales policies

Following the 1994 general elections South Africa returned to the international community. Despite the initial goodwill towards post-apartheid South Africa, one area of its foreign and security policy which has remained under close international scrutiny has been its arms trade policies. During apartheid the South

---

[55] Meldrum (note 54).

[56] Nzo, A., Statement before the Portfolio Committee on Foreign Affairs, Cape Town, 14 Mar. 1995, p. 11.

African arms industry operated in the international arms market in a highly clandestine manner. National legislation ensured that its activities, particularly arms exports, remained outside the realm of public scrutiny.[57] Over time a covert culture evolved within Armscor which was deep-rooted and insidious. Arms were exported to pariah states such as Chile, Iran, Iraq and Taiwan and to rebel groups such as UNITA and the Khmers Rouges.

With the advent of the government of President Nelson Mandela it was assumed that a new moral order, based on an ethical commitment to human rights, would help curtail the past excesses of the arms industry and South Africa's arms trade. Indeed, numerous statements from senior cabinet members have stressed the importance of adopting a moral and ethical approach to the arms trade. For instance, in November 1995 Deputy Prime Minister Thabo Mbeki told a meeting between the government and Armscor that the arms trade could not turn a blind eye to human-rights abuses.[58] Likewise, writing in the *Cape Times*, Professor Kader Asmal, Chairman of the National Conventional Arms Control Committee (NCACC), argued that 'moral, political and legal responsibility each requires that we grapple with, rather than wash our hands of, the hard issues of war and arms production'.[59] Putting such humanitarian principles into practice and overcoming the legacies of the previous regime have, however, proved an onerous task for the Government of National Unity, as is evidenced by the frequent outbreaks of scandal concerning illicit transfers of arms to less than reputable regimes.

In an attempt to earn the status of a responsible producer and supplier of weapons the Government of National Unity addressed the task of reforming its arms trade policies through a number of legislative measures designed to improve South Africa's arms control procedures and practices. A first priority was to earn international legitimacy by conforming to international norms and practices in arms control. In preparation for full membership of the Missile Technology Control Regime (MTCR) the government modified its national export regulations in the Non-proliferation of Weapons of Mass Destruction Act (Act no. 87) to include missiles and related dual-use technologies. The government also announced a ban on the production and supply of all land-mines.[60] On 13 May 1994, in Government Notice no. R88, issued by the Department of Defence, the government introduced licensing requirements for all items which fall within the limitations of the MTCR.[61] In October 1994 South Africa and the United States signed a bilateral missile-related import–export agreement which restated South Africa's intention to conform to MTCR

---

[57] Secrecy was institutionalized by the South African state through legislative measures such as the 1968 Armaments Development and Production Act, no. 1674 (as amended) which prohibited the disclosure of any information regarding the acquisition, supply, marketing, import, export, development, manufacture, maintenance or repair of or research on armaments, and the 1974 Special Defence Account Act which ensured that the facts behind South Africa's overseas sales were hidden from public scrutiny.

[58] Reuter, 'Arms trade needs ethics', *E. P. Herald*, 20 Nov. 1995.

[59] Asmal, K., 'The shape of SA's arms policy', *Cape Times*, 13 Nov. 1995.

[60] *Africa Research Bulletin*, 16 July 1994, p. 11804.

[61] *Government Gazette*, vol. 347, no. 15720 (13 May 1994).

guidelines and included provisions whereby South Africa could import space-launched vehicles to put its Greensat satellite into orbit. The two countries issued a joint statement outlining measures that South Africa would take to terminate its research programme on the development of space-launch vehicles.[62] In 1995 it joined the MTCR and the Nuclear Suppliers Group.[63] The government is also considering joining the Wassenaar Arrangement on Export Controls for Conventional Arms and Dual-use Goods and Technologies.[64] Such progress in conforming to international norms and values was marred by the failure of the government to submit details of arms transfers to the United Nations Register of Conventional Arms in the first two years of the new administration.

On the domestic front great emphasis has been placed on developing more open and transparent arms transfer policies. In October 1994 a commission of inquiry (the Cameron Commission) was set up to investigate all Armscor transactions after 1991, following press allegations of continued illicit arms transfers.[65] The commission made a number of recommendations for improving South Africa's arms trade policies. In August 1995 the Government of National Unity announced the launch of the NCACC and its guidelines and principles.[66] The Principles Governing National Arms Control stress that in pursuit of the general aim of an international cooperative and common approach to security the government will promote an effective arms control system and exercise due restraint in the transfer of conventional arms and related technologies by taking the following issues into account: (*a*) respect for the human rights and fundamental freedoms of the recipient country; (*b*) an evaluation based on the UN Universal Declaration of Human Rights and the African Charter on Human and Peoples' Rights (due consideration will be given especially to cases where political, social, cultural, religious and legal rights are seriously and systematically violated by the authorities of the recipient country); (*c*) the recipient country's internal and regional security situation in the light of existing tensions or armed conflicts; (*d*) the recipient country's record of compliance with international arms control agreements and treaties; (*e*) the nature and costs of the arms transferred in relation to the circumstances of the recipient country, including its legitimate security and defence needs and the objective of the least diversion of human and economic resources for armaments; and (*f*) the degree

---

[62] *US Department of State Dispatch*, vol. 5, no. 42 (17 Oct. 1994), p. 694.

[63] Also known as the London Club, the NSG coordinates multilateral export controls on nuclear materials and in 1997 agreed the Guidelines for Nuclear Transfers (the London Guidelines, revised in 1993).

[64] On the Wassenaar Arrangement, see Anthony, I. and Stock, T., 'Multilateral military-related export control measures', *SIPRI Yearbook 1996: Armaments, Disarmament and International Security* (Oxford University Press: Oxford, 1996), pp. 542–45.

[65] The most notorious case involved a shipment of arms to Yemen and the supply by Armscor of 2.4 million rand-worth of small arms and ammunition. Yemen was in a state of civil war at the time and subject to a UN arms embargo. The outcry led President Mandela to set up the Cameron Commission. It unearthed evidence of further illicit transfers to Angola, Rwanda and the former Yugoslavia. Brummer, S., 'SA's arms dealing underworld', *Mail and Guardian*, 2 June 1996, p. 9.

[66] South African Cabinet Office, Introduction to Press Conference: NCACC Rationale and Principles, Press release, Pretoria, 30 Aug. 1995. See also chapter 7, section IV in this volume.

to which arms sales are supportive of South Africa's national and foreign interests.[67]

Despite the substantial reforms introduced since 1994, South Africa's arms trade policy remains the subject of national and international controversy.[68] Conscious of the difficulties of establishing an arms trade policy determined by humanitarian principles, Asmal has stated: 'Hard issues arise for progressive policy makers in regulating the arms trade because industrial mechanisms are necessary to sustain military readiness. Because of this there is an ever-present risk that the military-industrial machine will overwhelm the moral compass that ought to guide it. The tail might wag the dog'.[69] Here he refers to arguments put forward by industry that if it is to survive then it must be allowed to trade with whomsoever it can secure markets with, thus 'erasing exactly the moral, political—sometimes even legal—concerns that should be paramount'.[70]

The lifting of the UN arms embargo in May 1994 and the decline in the defence budget, particularly the procurement budget, since 1989 created the incentives for the South African arms industry to maximize arms exports. Armscor, spearheading the drive for new overseas markets, stressed the positive macroeconomic benefits of increasing arms sales. Such arguments won over key elements of the formerly critical ANC. Even President Mandela, once a voice of opposition to the arms trade, appears convinced of the economic benefits of the arms trade. In a television address to the nation shortly after taking office he is recorded to have said: 'I don't think it would be fair to say that a particular country should not engage in trade in arms. Arms are for the purpose of defending sovereignty and the integrity of a country. From that angle there is nothing wrong in having a trade in arms'.[71] One prominent South African defence analyst commented on this apparent U-turn:

Since coming to power the ANC has changed dramatically from its pre-election stance. They have dropped their idealism and suddenly woken up to the fact that there is money to be made from the arms industry . . . For the ANC now facing the realities of government it is going to be very difficult to ignore the persuasive arguments by both local and international arms dealers that even though it's a dirty business it is potentially highly lucrative and the ripple effect on the economy is tremendous.[72]

Joe Modise, the new Minister of Defence, is particularly keen to promote a flourishing arms industry even when this means exporting arms to less than reputable regimes.[73] Modise emphasizes the role of the national arms industry as an essential attribute of a strong sovereign state, which helps to bestow

---

[67] South African Cabinet Office (note 65).
[68] 'South Africa: Damascus deal', *Oxford Analytica Daily Brief*, 22 Jan. 1997, p. 13.
[69] Asmal (note 59).
[70] Asmal (note 59).
[71] Quoted in *Southscan*, 27 May 1994.
[72] Cilliers, J., *South African Sunday Times*, 12 July 1994.
[73] Joe Modise's views were captured in an interview by Helmoed Romer Heitman in *Jane's Defence Weekly*, 6 Aug. 1994.

power and influence on South Africa, both within the Southern African region and further afield.[74]

Maintaining a sizeable arms industry is not necessarily incompatible with a commitment to human rights, but as a 'third-tier' arms supplier[75] South Africa has found it difficult to break into world markets, dominated as they are by the major arms suppliers. This gives South African firms little option but to supply to regimes that do not always conform to international norms and standards, particularly in the field of human rights.[76]

In pursuit of new markets the industry has placed a strong emphasis on international joint ventures and strategic alliances and on arms exports to the rest of Africa, in particular to Southern Africa, which is perceived as South Africa's natural sphere of influence. The argument, according to various commentators, is that selling arms will 'help allies and friends in the region deter or defend themselves against aggression, while promoting interoperability with South African forces when combined operations are required'.[77] In this manner arms sales to the region are promoted by defence officials as a key to establishing regional security cooperation to the benefit of all countries within the region.[78] The South African arms lobby has been quick to point out that dependence on the major (i.e., Western) military suppliers for military equipment could compromise the national security policies of African recipients. Such altruism is hardly the reason for promoting arms sales in the region. Pierre Steyn, the Secretary for Defence, has argued that:

Judicious arms sales can prove valuable both economically and politically in the sphere of foreign relations. Especially in Africa, South African weapon sales may expand South African influence across the continent. This may also benefit African states in their search to lessen dependence upon non-African supply sources, moving away from North American and European suppliers to South Africa. Money spent would remain in Africa.[79]

The purported economic benefits of regional arms sales take the form of enhanced economies of scale for the South African arms industry. An additional benefit, often not publicly voiced, is that arms sales to the region will in all likelihood lock recipients into a relationship of dependence on South Africa's arms industry through spares and maintenance agreements. As Steyn implies, to succeed in becoming the principal regional supplier would bestow on South Africa a considerable amount of political leverage over regional states.

[74] Steyn, P., 'Aligning the South African defence industry with regional integration and security co-operation', *African Security Review*, vol. 4, no. 2 (1995), pp. 15–23.
[75] Anthony, I., 'The third-tier countries: production of major weapons', ed. H. Wulf, SIPRI, *Arms Industry Limited* (Oxford University Press: Oxford, 1993).
[76] See chapter 7 in this volume for details of arms scandals since 1994.
[77] Cilliers, J., 'Towards a South African conventional arms trade policy', *African Security Review*, vol. 4, no. 4 (1995), pp. 3–20.
[78] Steyn (note 74).
[79] Steyn (note 74), p. 19.

Those wanting to curtail South Africa's arms trade have expressed concern that the promotion of arms sales to the region would divert scarce resources to military ends, thereby threatening the prospects for regional development.[80] Cock, for instance, argues that regional security would be far better served if South Africa desisted from arms sales and concentrated instead on promoting sustainable development through non-military forms of cooperation.[81] According to certain defence analysts the current policy of developing bigger regional markets for South African arms 'is more likely to lead to greater insecurity and heighten conflict potential' than to make a positive contribution to a new regional security order, as its advocates claim.[82] The potential for South Africa to dominate the region politically, through its military–industrial potential and economic strength, is a growing cause for concern.[83] As a consequence most countries in the region remain hesitant about procuring arms from South Africa, not wanting to increase South Africa's political leverage over them.[84]

While consciously avoiding buying armaments from South Africa, many countries in the region, including Angola, Botswana, Namibia and Zimbabwe, have been involved in procuring significant amounts of weapons from foreign suppliers, thus reversing the declining trend in military expenditures in Southern Africa since roughly 1989. The large number of arms purchased by various countries in the past few years has lead to increasing tensions within the region and fears of a regional arms build-up.[85]

Recent weapon imports into Angola have included small arms, ammunition, tanks, attack helicopters and fighter aircraft from numerous sources, including the governments of Brazil, Bulgaria, the Czech Republic, Poland, Portugal, Russia and Ukraine. In 1995 and 1996 Botswana embarked on an ambitious programme of arms purchases including orders for 36 Scorpion light tanks from Britain and combat aircraft (F-5s) from Canada.[86] These trends have raised alarm among neighbouring countries, particularly in Namibia, which has a territorial dispute with Botswana over the sovereignty of Sedudu Island (Kasikili Island to the Namibians). In response to these trends Namibia has increased its defence budget in the past two years in order to purchase Russian equipment.[87]

In justifying the growth in military capabilities Botswana's military leaders explain that they need to be ready for any eventuality and that the current peaceful situation in Southern Africa may not endure.[88] Perhaps more convincingly, they argue that the Botswana Army needs to equip itself for peace-

[80] George (note 49); and *Human Development Report 1994* (note 35).

[81] Cock, J., 'Rethinking security, the military and the ecology of Southern Africa', Group for Environmental Monitoring Discussion Document, Johannesburg, Mar. 1995.

[82] George (note 49), p. 262.

[83] George (note 49), p. 255.

[84] This view is based on private communications between the authors and senior military officers from various SADC countries.

[85] 'Hawkish Botswana', *The Star and SA Times International*, 19 June 1996, p. 23.

[86] Rakabane, M., 'Bizarre arms race', *New Africa*, Jan. 1997. An earlier attempt to buy 50 second-hand Leopard tanks and 200 army trucks from the Netherlands was blocked by Germany which found itself under intense lobbying pressure from Namibia.

[87] 'Botswana splashes out on arms', *Mail and Guardian*, 4 Apr. 1996.

[88] Rakabane (note 86).

keeping operations given its recent involvement in peacekeeping in Mozambique and Somalia. Peacekeeping does require a considerable amount of equipment, particularly for logistical support, but it is questionable whether this includes tanks and fighter aircraft. A more likely explanation for Botswana's military build-up is that many years of dynamic economic growth have enabled its leaders to purchase the prestige symbols of an emerging sovereign state. Whatever the reasons for Botswana's purchases, its arms build-up goes against the current trend of disarmament and security cooperation in the region.

While Namibia may have genuine cause for concern over the implications of Botswana's enhanced military capability, South Africa's reaction to it has been somewhat alarmist.[89] Botswana's military acquisitions pale into insignificance beside South Africa's planned purchases of corvettes, combat aircraft and submarines.[90]

The suspicions and tensions reflected in the press over certain arms purchases in the region suggest that so far there has been only limited success in building regional confidence and security in the area of arms acquisitions and control and that far more needs to be done at the national and regional level if arms purchases are not to contribute to regional tensions.

## VI. Strategic doctrine

The broadening of South Africa's security agenda has clearly encouraged the South African military to redefine its role. General George Meiring, Chief of the National Defence, has stated:

Our own region is not very stable and extremely vulnerable against dormant accelerators such as drought, over-population, poverty, famine, endemic diseases, illiteracy and others, which, if not managed correctly will lead to a serious instability in the Southern African region, with disastrous consequences. In such a scenario there can be no developments or improvement in the quality of life for our people. It is therefore in the Republic of South Africa's economic and security interests that, in a world where the lamb and the lion are ostensibly preparing to lie down next to one another, the Republic of South Africa will have to be the lion.[91]

Essentially the military argues that the new security challenges can only be met by establishing stability through a strong military balance of power. This would appear to contradict the new principles of security elaborated in the 1996 White Paper on Defence, but a close reading of that document exposes an essential contradiction between the new thinking on security and the persisting traditional military approaches. Old thinking is evident in the statements: 'The SANDF may be employed in a range of secondary roles as prescribed by law, but its primary and essential function is service in defence of South Africa, for

---

[89] *Mail and Guardian* (note 87).
[90] 'South Africa favours UK arms deal', *Sunday Times*, 30 Mar. 1997.
[91] South African Department of Defence, *The National Defence Force in Transition: Annual Report Financial Year 1994/95*, p. 1.

the protection of its sovereignty and territorial integrity'; and the 'SANDF shall be a balanced, modern, affordable and technologically advanced military force, capable of executing its tasks effectively and efficiently'.[92]

The essential contradictions between the traditional and new approaches to security which exist side by side in the White Paper have come about as a result of a process of compromise between the old and the new, which is an essential feature of the 'negotiated revolution'.[93] The document was written following a long process of consultation which encouraged submissions from a broad spectrum of groups in society, including the military.[94] While the openness, transparency and principles of democratic consultation of the Defence Review process are highly commendable, they do not necessarily produce coherent strategy. Rather they capture the tensions and contradictions which still exist within South Africa's security community. Nowhere are these contradictions more apparent than in the attempt to formulate a new strategic doctrine.

The White Paper on Defence states that 'the SANDF shall have a primarily defensive orientation and posture'.[95] Article 227 of the interim constitution states that 'the national defence force shall be primarily defensive in the exercise and performance of its powers and functions'.[96] This is only slightly less unequivocal than the statement in the ANC Policy Guidelines that 'the Defence Force shall be defensive in its character, orientation and its strategy, and its force level will be adjusted accordingly'.[97]

A shift towards a more defensive strategic posture will have significant implications for the types of weapons deployed by the SANDF. The stakes in this debate have therefore been high for the arms industry. Not surprisingly, arms industry organizations such as SADIA and Armscor have played a proactive role in the debates surrounding the 1996 White Paper and the Defence Review.

The 'offensive defence' posture—a strategy which is defensive, but a posture which is offensive—which the SADF adopted during the apartheid era has remained at the core of the SANDF's strategic doctrine since the change of government in 1994.[98] The idea behind this posture was that the military must be prepared at all times to protect South Africa's territorial integrity by taking offensive, proactive steps to avoid combat on South African soil. The doctrine

[92] White Paper on National Defence (note 5), p. 6.

[93] Ohlson, T., 'South Africa: from apartheid to multi-party democracy', SIPRI Yearbook 1995: Armaments, Disarmament and International Security (Oxford University Press: Oxford, 1995), p. 117.

[94] Interview with Laurie Nathan, drafter of the White Paper, Centre for Conflict Resolution, Cape Town, 14 Feb. 1996.

[95] White Paper on National Defence (note 5), p. 6.

[96] Republic of South Africa, Constitution of the Republic of South Africa: Act no. 200 of 1993, Government Gazette, no. 15466 (28 Jan. 1993).

[97] African National Congress, ANC Policy Guidelines for a Democratic South Africa, adopted at the ANC National Conference, 28–31 May 1992, p. 3 (section 2.12).

[98] For a detailed discussion of the debate on strategic doctrine, see Siko, M. and Cawthra, G., 'South Africa: prospects for NOD in the context of collective regional security', NoD and Conversion, no. 33 (July 1995).

seeks to deter external aggression by threatening to inflict high levels of destruction on the enemy via pre-emptive or retaliatory offensive action.[99]

The dominant concerns of South Africa's strategic planners during the apartheid era related to questions about regional strategic balance, whether the enemy threat was growing, and whether the deployment of a certain weapon system would maintain stability and/or provide military superiority. Most of these questions were concerned with the deployments of weapon systems. This obsession leads to what some analysts have described as a 'weapons culture'.[100]

In the Southern African context, given the vast distances, relatively flat terrain and underdeveloped infrastructure in most of the states in the region, the SANDF's offensive defence posture requires good logistical support and air transport systems in conjunction with paratroopers, allowing for rapid interdiction and supply over long distances. It also requires a strong, versatile helicopter capability—the backbone of search-and-rescue missions—highly mobile armoured vehicles like the Ratel and the Rooikat, and fire support systems such as the G5 and G6 artillery systems. Air force equipment includes strike interceptor aircraft and an in-flight refuelling capability. The offensive nature of the structure of the SANDF is encapsulated in the Special Forces Regiment, military training practices, and the logistical and communications infrastructure which is designed to support offensive operations over long distances.[101]

The SADF's doctrine of offensive defence was inherently provocative and destabilizing within the Southern African region and is no less so in the current situation.[102] When a state considers pre-emptive strikes and surprise attacks as a strategic option it effectively puts pressure on neighbouring states to emphasize the same strategies. The all-too-likely result is an escalating arms build-up. This is the classic action–reaction spiral or what is often termed the 'security dilemma'—the vertical and horizontal proliferation of arms in response to the real or perceived threat from one's neighbours. In these conditions the risk of armed conflict is amplified and the security of all states in the region reduced.

The present, more benign, regional climate has seriously eroded the foundations upon which the SADF's traditional strategic doctrine was constructed. The time-honoured constructs of threat and deterrence no longer hold weight in a region undergoing a perceptible process of disarmament and demobilization.[103] Nor do they provide justification for the retention of strong military forces. This benign security environment coincides with bilateral and multilateral donor pressure for disarmament, demobilization and demilitarization

---

[99] Kruys, J., 'The defence posture of the SADF in the nineties: some geo-strategic determining factors', Paper presented to the Institute of Strategic Studies Conference, University of Pretoria, Pretoria, 25 June 1992.

[100] Booth, K., 'Strategy', eds A. Groom and M. Light, *Contemporary International Relations: A Guide to Theory* (Pinter: London, 1994), p. 316.

[101] Siko and Cawthra (note 98), p. 21.

[102] Nathan (note 7), p. 75.

[103] Willett, S., *Military Spending Trends and Development Cooperation in Southern Africa: South Africa, Angola, Mozambique and Zimbabwe* (OECD, Development Assistance Committee: Paris, Jan. 1997).

within the region.[104] The redundancy of offensive doctrines is reflected in Booth's observation that 'states are running out of motives for war'.[105]

Linked to these trends is a growing recognition of the way in which arms purchases in the region, far from producing stability and security, have in fact increased the levels of destructive power and insecurity (the security dilemma) and placed a heavy burden on the Southern African economies.[106] As a consequence South Africa's new security advisers have sought to secure a strategic doctrine that is more in keeping with the atmosphere of détente and cooperation within the region. This has led to the promotion of non-offensive defence (NOD) as a means of providing an operational doctrine to the concept of defensive defence.[107]

The basic premise of NOD is the establishment of a security regime within which domestic, regional and international peace and security can be guaranteed. A central tenet of this doctrine is to combine security assurances with peaceful intentions. This implies restraint from the threat or use of force against the territorial integrity or political independence of any state. These notions are to be found in the UN literature on defensive defence which describes 'a condition of peace and security attained step by step and sustained through effective and concrete measures in the political and military fields'.[108]

NOD is also conceptualized as 'a strategy, materialised in a national posture that emphasises defensive at the expense of offensive military options'.[109] It rejects the idea that military stability is achieved through a balance of force, and argues instead that security is achieved through the non-threatening behaviour of states. Underlying this approach is a profoundly different way of conceptualizing military strategy. Instead of being preoccupied with real or perceived external threats, it is concerned with the perceptions of other states over its own projection of power, that is, whether it is perceived as benign or offensive. Replacing an offensive doctrine with one of non-offensive defence would help to bring South Africa's military doctrine into line with its broader political aims of regional cooperation and confidence and security building.

NOD actively seeks to reduce the arms race, and the underlying insecurities associated with it, by refraining from the classical regional security dilemma whereby the security of South Africa is achieved at the expense of its neighbours.[110] It is based on the notion of defensive superiority: the defensive capabilities of a country render the prospect of successful attack both unattractive

[104] Solomon and Cilliers (note 48).
[105] Booth (note 100), p. 316.
[106] George (note 49), pp. 262–63, 266–68.
[107] For a detailed discussion of this concept, see Boderup, A. and Nield, R., *The Foundations of Defensive Defence* (Macmillan: London, 1990); Conetta, C., Knight, C. and Unterseher, L., 'Towards defensive restructuring in the Middle East', *Bulletin of Peace Proposals*, vol. 22, no. 2, pp. 115–34; and Möller, B., *Non-Offensive Defence as a Strategy for Small States*, Working Paper no. 5 (Centre for Peace and Conflict Research: Copenhagen, 1994).
[108] UN Office for Disarmament Affairs, Study on Defensive Security Concepts and Policies, Disarmament Study Series, no. 26, UN document A/47/394, 1993, section 87.
[109] Möller (note 107).
[110] Nathan (note 7), p. 78.

and unattainable. It seeks to wear down an invader through the combination of attrition and highly offensive action using killing fields, tank traps, anti-tank weapons, minefields, light infantry, mobile armour for counter-attack, and the destruction of enemy aircraft through anti-air systems and defensive air attacks.

For a developing country such as South Africa facing considerable resource constraints, NOD is relatively cheap because of the savings derived from the fact that certain expensive systems required for offensive defence can be dispensed with, such as command, control, communications and intelligence ($C^3I$) systems, offensive air and missile capabilities and logistics systems. Moreover, fewer forces are required by a defender to repulse a hostile force. The usual yardstick for the ratio of attacking to defending forces if attack is to succeed is thought to be 3 : 1. The cost savings implied by NOD make it extremely attractive to post-apartheid South Africa where the socio-economic improvement of the country's poor and marginalized communities is one of the key national priorities. In strategic terms there are advantages because of the ability of the defender to control the terms of the battle and because the defender knows the terrain. Moreover, the defender can take advantage of force multipliers such as dispersed structures and formations. Finally, a defender will tend to have superior moral support from both its domestic population and the international community.

## VII. The military backlash

Non-offensive defence has not found much favour with the South African military establishment as it constitutes a major challenge to the SANDF's existing operational practice and force structures. The perceived challenge to its traditional modus operandi has put the SANDF on the offensive in the debates on strategic doctrine which have taken place during the Defence Review.

The SANDF has attempted to identify a host of new threats requiring military responses. These include the collapse of South Africa's neighbours, resulting in regional instability, and a united and coordinated attack by its neighbours, possibly orchestrated by a foreign superpower.[111] On the first scenario, General Meiring has said: 'Overflow from regional conflicts is the most likely threat to our national security, mainly in two ways: a large scale influx of illegal immigrants and refugees from countries in the region; the RSA inadvertently being drawn into regional conflict'.[112] The relationship between migration and security has become a major concern in the Southern African security community.[113]

Meiring also insists that 'in an uncertain world it must be assumed that the enemy will be strong and sophisticated'.[114] It follows that the South African

[111] These senarios were presented by Gen. George Meiring at the ANC/SADF Bilateral Expenditure Talks, Pretoria, Mar. 1994.
[112] Meiring, G., 'Taking the South African Army into the future', *African Defence Review*, no. 14 (Jan. 1994), p. 2.
[113] Christie, K., 'Security and forced migration concerns in South Africa', *African Security Review*, vol. 6, no. 1 (1997).
[114] Meiring (note 112), p. 3.

Army needs to improve its fighting quality through investment in: (*a*) research and development; (*b*) technology, including aspects such as defensive chemical warfare capability; (*c*) war gaming and operational research; and (*d*) force multipliers such as electronic warfare and night fighting capability and the mobility, reach and staying power of its forces.[115]

The SANDF has reorganized itself around the principles of a threat-independent approach and a core-force approach. The threat-independent approach lays stress on the present geo-strategic uncertainties and the consequent need for a wide range of responses given all possible contingencies. The core-force approach relies on the idea of a five-year warning period for any new conventional threat, allowing the SANDF the time to build itself up to a full war mobilization capability.[116]

The primary orientation of this structure and strategy continues to emphasize the external strategic role of the armed forces when in reality the SANDF, in common with most African armies, is principally engaged in internal security operations. For instance, during the election period the SADF deployed nearly 20 000 troops for internal security purposes.[117] The deployment of troops in support of the police (including border patrol) is highly unpopular with the SANDF, but how long this role lasts will be determined by the persistence of internal dissent and rising violence.[118] In an ideal world it should be possible to separate internal and external security roles by allocating the former to the South African Police and allowing the armed forces to concentrate their efforts on external security threats. In reality the highly politicized and poorly trained police service is unable to render an effective service, either in terms of tackling crime or in handling internal political violence. This has meant that the SANDF has had to be deployed in support of the police in certain politically sensitive areas such as the East Rand and KwaZulu–Natal and in more conventional policing roles in the urban areas. Given the poor conditions of work and low levels of training and morale in the police force it is likely to be some time before a professional force emerges which commands respect and legitimacy from the population at large. Thus it seems that the SANDF's internal security role will continue for the foreseeable future.

Despite the fact that the SANDF's secondary roles have been increasing in importance since 1994, the notion of a classical defence force configured to protect the country against external conventional military threats continues to dominate the thinking of the defence force planners and strategists. As Williams notes, however, 'the determinants of roles, missions and tasks, and the planning and design of the armed forces in these contingencies (internal deployments) is a very different conceptual and practical exercise to planning for classical threats'.[119] Some fundamental contradictions exist between the policy of

---

[115] Meiring (note 112), pp. 3–4.
[116] Draft Defence Review Document, 15 Feb. 1997 (note 9).
[117] *The National Defence Force in Transition* (note 91), p. 67.
[118] White Paper on National Defence (note 5), pp. 28–29.
[119] Williams (note 11), p. 31.

defensive defence enshrined in the 1996 White Paper and the largely offensive defence posture approved for the SANDF in the 1997 Defence Review. These contradictions are likely to be heightened in the coming years.

## VIII. Implications for the arms industry

In his *Introduction to Strategic Studies*, Buzan has defined strategic planning as the military technological variable within international relations.[120] This sense of strategic doctrine as a technological variable reveals the crucial link between the arms industry and South Africa's security function. Strategic doctrine defines force structure, deployment and weapon capabilities and thus informs and structures defence procurement decision making. This section briefly examines the procurement and therefore the defence industrial implications of the strategic doctrines outlined above.

The implications for South Africa adopting a non-offensive defence posture would be the 'restructuring [of] every aspect of the armed services—posture, development, weaponry, training manpower, procurement etc., so as to bring the overall character and strategy of the military into line with declared defensive intentions'.[121] As a general rule, a defensively-oriented force can make greater use of simpler technologies because it benefits from operating on (or over) familiar and prepared territory. By contrast the challenge of projecting soldiers and guiding weapons deep into enemy territory can compel a reliance on more costly technologies. Transforming the SANDF from its present offensive configuration into one that adopted a genuinely defensive posture in the form of an operational doctrine based on the principles of non-offensive defence would require a major restructuring of its present structures and practices. In summing up the implications of a defensive posture, Williams argues that it 'is an all encompassing concept which addresses all aspects of defence policy including weapons innovation and the procurement process: force structure planning and design: domestic defence posture and doctrine: arms controls and arms exports, and force application and employment'.[122]

Restructuring on these lines would have enormous implications for the South African arms industry. First, it would involve the cancellation of many existing weapon programmes which are essentially offensive in nature, such as the Olifant main battle tank and the Rooivalk attack helicopter. It might also imply the shelving of plans to purchase such items as submarines, corvettes and frontline combat aircraft. This latter would have less impact on the domestic arms industry as these items would be bought off the shelf from foreign suppliers: although considerable offsets which would be attached to these foreign purchases would be lost (local suppliers and subcontractors would expect to pick

---

[120] Buzan B., *An Introduction to Strategic Studies: Military Technology and International Relations* (Macmillan in association with the International Institute for Strategic Studies: Basingstoke, 1987).
[121] Nathan, L., 'Towards a post-apartheid threat analysis', *Strategic Review for Southern Africa*, vol. 15, no. 1 (1993), pp. 43–71.
[122] Williams (note 11), p. 52.

up business from the offset deals linked to these foreign purchases), the cancellation of such programmes would have little significant effect on the size and structure of the existing South African defence industrial base. However, the cancellation of other offensive systems, such as main battle tanks, attack helicopters and long-range missiles, which are built in South Africa, would have a negative impact on local defence firms such as Denel and Reunert. On the other hand the SANDF's potential involvement in peace support operations and its present commitment to mine-clearance programmes could well stimulate demand in other sectors of the defence market, for instance, for armoured personnel carriers (APCs), transport helicopters, mine-clearance equipment and mine-protected equipment.

A second factor which would affect the domestic arms industry is the creation of a link between defence posture and arms sales. While an offensive posture might encourage arms exports in order to gain better value for money from domestic procurement by enhancing economies of scale, a defensive doctrine which stresses the role of security cooperation and conflict prevention is likely to curtail arms transfers and further reduce defence industrial activities as part of a regional confidence-building programme. However, if the South African arms industry manages to persuade its neighbours of the advantages of standardization, thus encouraging them to purchase South African defence equipment, then the arms industry might be prevented from down-sizing any further. Rather the patterns of production will switch in accordance with the shifting patterns of demand. In addition, systems suitable for peacekeeping and cooperative regional operations are likely to be far less controversial in terms of South Africa's new arms export policy guidelines. There would also probably be less public opposition both inside and outside South Africa to such arms transfers.

The SANDF's proposed core-force concept supposes a balanced force retaining key across-the-board capabilities and includes a rapid-deployment force comprising elements of all the arms of service. The key capabilities include strategic intelligence, counter-air, maritime strike, strategic interdiction, tactical air support, air reconnaissance, mobile ground forces, rear area defence, $C^3I$ and logistical support capabilities.[123] The implication of these capabilities is that South Africa will need to invest in new offensive equipment in the near future in order to maintain its present level of competence in all the existing arms of service. Hence the existence of a procurement 'wish list' which includes attack helicopters, main battle tanks, combat aircraft, corvettes, submarines and other major offensive weapon systems.

In contrast to the primary military functions of the core-force concept, the secondary functions of internal security and peacekeeping require quite different configurations of manpower and equipment. The emphasis in both peacekeeping and internal security operations is on manpower, logistics and support

---

[123] Heitman, H., 'Reshaping South Africa's armed forces', in 'Whither South Africa's warriors?', *Jane's Intelligence Review*, Special Report no. 3 (1994).

equipment, such as transport aircraft and troop-carrying and reconnaissance helicopters, APCs, trucks and jeeps rather than on highly offensive front-line systems such as combat aircraft, tanks and missile systems. Strategic planners in South Africa are aware of the tension which exists between the likelihood of the SANDF performing its primary function and the demands of its secondary functions.[124] The orthodox response to this situation is to design, budget and procure for the primary function and perform the secondary functions with the collateral utility derived from the primary force design.

This attempt to resolve the problem is becoming less and less tenable in a situation of budgetary constraints and increasing political opposition to an enhanced role for the SANDF in the new South Africa. Moreover, parliament and the treasury have become increasingly reticent about supporting the allocation of scarce resources to the SANDF's primary functions in the light of the pressing need for it to execute its more immediate and politically pressing secondary roles effectively.[125] Williams suggests that a pragmatic solution would be to plan, design and budget for the SANDF to carry out its secondary roles. This would involve designing multi-functional forces capable of executing a wide range of roles. It would also, more importantly, require a fundamental transformation of the SANDF's existing force structures and equipment deployments.[126]

## IX. A historic compromise

The challenge facing the recent Defence Review process was to translate the policies of the White Paper into operational practice. This represents the greatest test for the forces of change within the government and the military establishment. The focus on a defensive force structure and force design has generated much opposition from within the SANDF, particularly as certain advocates of non-offensive defence have proposed the scrapping of tank regiments, the disbanding of combat aircraft squadrons and the denial of corvettes and submarines.[127] Cutting such technologies goes to the very heart of service structures and the way the military sees itself, hitting the cultural nerve of militarism, its traditions and its practices. In other countries the experience of defence reviews indicates that radical proposals for restructuring and change become watered down as a result of renegotiation through the 'back door', that is, via the executive and as a result of the effective barriers to implementation that the defence forces invariably construct.

---

[124] Interview with Capt. Fanie Uys, strategic planner in the Department of Defence, Cape Town, Feb. 1996.

[125] Interview with Dr Ian Phillips, ANC MP and member of the Joint Standing Committee on Defence, Cape Town, Feb. 1996.

[126] Williams (note 11), p. 52. See also Cilliers, J., 'Defence research and development in South Africa: the role of the CSIR', *African Security Review*, vol. 5, no. 5 (1996), p. 39.

[127] Interviews with Laurie Nathan, drafter of the Defence White Paper, and Dr Ian Phillips, Cape Town, Feb. 1996.

Other factors have also tempered the adoption of a genuine non-offensive defence policy. Neighbouring countries have made repeated requests for assistance in various defence-related fields such as maritime assistance.[128] Responding to these and other commitments would require the maintenance of certain capabilities such as corvettes, with range and flexibility, which in certain circumstances can be used in offensive roles. A similar observation applies to South Africa's possible involvement in peacekeeping operations which would require logistical support systems with range and manoeuvrability.

The fact that questions of policy, doctrine and practice are under substantial review does not mean that the SANDF will be totally restructured along the lines implicated by the policy framework of the White Paper. It seems likely that there will be no clear winner, rather a compromise between the 'doves' and the 'hawks', resulting in a less militaristic stance than under apartheid, but nevertheless retaining a relatively strong conventional military force with some offensive capabilities.[129] What has been significant in the Defence Review process, however, is that the power to formulate policy is being quite rapidly shifted from the SANDF to the Ministry of Defence, the civilian-controlled Defence Secretariat and, to a limited but nevertheless significant extent, to civil society via the Defence Select Committee.

The current debates on arms trade policies and control, strategic doctrine and security policy which have been examined in this chapter are relevant to the future of South Africa's arms industry. The outcome of these debates will inform decisions concerning future procurement choices, which in turn will determine the future shape and size of the domestic arms industry. However, no automatic relationship can be assumed between a shift in strategic doctrine from offensive to defensive defence and a parallel shift in the size and structure of the South African arms industry.

---

[128] Interview by Helmoed Romer Heitman with Joe Modise, *Jane's Defence Weekly*, 3 Jan. 1996, p. 32.
[129] This compromise is apparent in the draft Defence Review documents (note 9).

# 7. Defence industrial adjustment after 1994

## I. Introduction

The changes in South Africa's security environment since the ending of apartheid have been accompanied by a number of political and institutional changes in the defence establishment. The most significant of these have included the restructuring of the Department of Defence, the establishment of a civilian Defence Secretariat and the formation of a parliamentary defence committee with constitutionally defined powers and functions. These changes have fundamentally altered the pattern of civil–military relations in favour of the civilian political authority, with the result that the SANDF has become increasingly marginalized with respect to decision making on defence matters.

The increasing prominence of civilians, including parliamentarians, in defence decision making has also had a significant impact on the local military–industrial complex. The domestic arms industry has been subject to much higher levels of public scrutiny than during the apartheid era, largely as a result of the work of the Cameron Commission[1] and the public policy processes associated with the 1996 White Paper on National Defence and the Defence Review. As a result of these developments the industry has been forced to become more transparent and accountable to the public and to parliament. In an attempt to maintain its influence over defence matters and to preserve its vested interests, the local military–industrial complex, led by Armscor, has become an enthusiastic supporter of the government's Reconstruction and Development Programme.[2]

Notwithstanding the concerted efforts of the local military–industrial complex to argue against further defence cuts, since April 1994 the defence budget has been reduced even further as a result of budgetary constraints and more pressing spending priorities. In this context, and in the absence of any clear government policy, the arms industry has been forced to continue its ad hoc and market-driven strategy of down-sizing and restructuring in order to survive the impact of the decline in the domestic defence market. The lifting of the UN arms embargo in May 1994 and South Africa's acceptance back into the international community have subjected the country's arms industry to higher levels of competition. They have also prompted the arms industry to pursue arms exports and international collaboration quite vigorously, with the result that the country's defence industrial base has become increasingly internationalized since the ending of apartheid.

This chapter examines the political and institutional changes and policy developments which have taken place in the defence establishment since April

---

[1] See section IV in this chapter; and note 11.
[2] *Armscor Annual Report 1995/96*, p. 35.

1994. It considers the impact of these changes on the South African defence industrial base and examines the adjustment experiences of the arms industry since April 1994.

## II. Political and institutional changes in the defence establishment

South Africa's defence establishment has witnessed a number of fundamental political and institutional changes since the country's first democratic elections.

Shortly after taking office in May 1995 the ANC-led Government of National Unity appointed a new minister and deputy minister of defence, Joe Modise and Ronnie Kasrils, respectively. The appointment of two 'hawkish' ANC ministers to the defence portfolio and the appointment of the former head of the SADF, General George Meiring, as chief of the SANDF for a period of five years from 27 April 1994 reflected the compromises achieved during the Joint Military Co-ordinating Council (JMCC) negotiation period. Meiring's appointment was politically highly significant because it was intended to placate reactionary elements within the SADF who remained opposed to the transition to democracy and to 'balance' the appointment of two ANC ministers to the defence portfolio.[3] The basic structure of the SADF continued into the post-election period, as agreed during the JMCC negotiations.[4] It consists of four services—army, air force, navy and medical services—and supporting staff departments, while the procurement function for the defence force remains vested in Armscor.

Despite the continuity in the organization and structure of the SANDF, the Department of Defence was extensively restructured in the months following the elections in order to reflect the principles of democratic civil–military relations and the supremacy of civilian authority over the armed forces contained in the provisions of the interim constitution.[5] Shortly after taking office the new minister initiated a process to establish a 'balanced-model' Ministry of Defence, which included the establishment of a civilian Defence Secretariat and the appointment of a Defence Secretary.[6] The organizational structure of the new Department of Defence is shown in figure 7.1.

The 'new' Ministry of Defence comprises the Office of the Minister of Defence, the Defence Secretariat, SANDF Headquarters and Armscor. The Department of Defence comprises the constituent parts of the Ministry of Defence and the SANDF. Under the balanced model the Secretary of Defence, who is head of the Defence Secretariat, is the accounting officer of the Depart-

---

[3] 'Whither South Africa's warriors?', *Jane's Intelligence Review*, Special Report no. 3, July 1994, p. 10.

[4] Williams, R., *South Africa's New Defence Force: Progress and Prospects*, CSIS Africa Notes no. 170 (Center for Strategic and International Studies: Washington, DC, Mar. 1995).

[5] Republic of South Africa, Constitution of the Republic of South Africa: Act no. 200 of 1993, *Government Gazette*, no. 15466 (28 Jan. 1993), sections 225–28.

[6] The Defence Secretariat was abolished in the late 1960s. The establishment of a 'new' civilian Defence Secretariat to enhance parliamentary control of the defence function was a reform target of the RDP. *SANDF Annual Report 1994/95* (Department of Defence: Pretoria, 1995), p. 42.

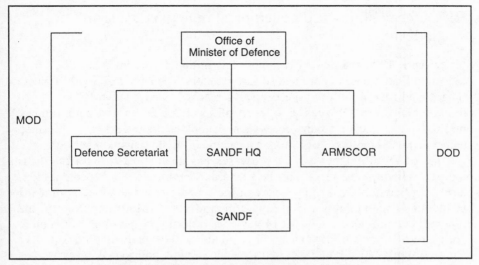

**Figure 7.1.** Department of Defence: organizational structure, 1995
*Source: SANDF Annual Report 1994/95* (Department of Defence: Pretoria, 1995), p. 42.

ment of Defence and the principal civilian adviser to the minister regarding defence policy matters. The Chief of the SANDF is the head of the Department of Defence and the principal adviser to the minister on military and operational matters.[7]

These political and institutional changes, particularly the establishment of the Defence Secretariat and a parliamentary defence committee, together with the absence of any government policy towards the arms industry, have placed the industry 'at arm's length' from the state and thereby contributed to the reduction of the influence of the apartheid-era military–industrial complex. At the same time as the distance between the arms industry and the state has grown, there has been a surprising convergence of interests and views between the ANC-dominated leadership of the Ministry of Defence, in particular the Defence Minister, Joe Modise, and former SADF senior military officers, including General Meiring.[8] This rapprochement between certain elements of the new and old political orders reflects the shifting alliances and accommodations which are a distinct feature of South Africa's transition to democracy, and which are reflected in the current debates on the future of its arms industry.[9]

[7] *SANDF Annual Report 1994/95* (note 6), pp. 42–43.
[8] Interviews with Joe Modise, *Jane's Defence Weekly*, 6 Aug. 1994; and 3 Jan. 1996.
[9] 'Modise highlights benefits of SA defence industry', *Business Day*, 14 Nov. 1996.

## III. Defence policy and government policy on the arms industry

Since April 1994 the new government has initiated a number of policy processes with regard to defence and security matters. The nature of South Africa's 'negotiated revolution' and the decisions reached during the JMCC negotiations, which were achieved as a result of accommodation and compromise, meant that the defence policy processes in the post-apartheid era were informed and constrained by these 'negotiation legacies' from the transition period.[10]

None of these defence policy processes has dealt directly with the arms industry, although the reports of the Cameron Commission have been particularly significant in challenging many of the economic and strategic arguments in favour of maintaining a domestic arms industry.[11] More importantly, the reports have provided a new set of values and norms for guiding government policy on the arms trade and the arms industry. The inclusive, consultative nature of the 1996 White Paper and Defence Review policy-making processes has, however, allowed many groups in civil society to put forward their views on the future of the domestic arms industry.

During the JMCC process in early 1994 a draft 'National Policy for the Defence Industry' was produced and presented to the TEC Sub-Council on Defence in April 1994. The main thrust of this document, which was approved by the Sub-Council on Defence, was preserving rather than restructuring the apartheid arms industry.[12] The document argued in favour of a policy of maximizing arms exports, subject to certain guidelines, in order to offset the decline in the domestic market. The recommendations have not been formally adopted by the new government but have been used to provide an informal policy framework for the arms industry since April 1994.

In August 1994 the cabinet set up a committee under the chairmanship of the Minister of Defence to develop a policy framework for dealing with the domestic arms industry. However, no concrete proposals emerged from the work of this committee because it was recognized that new defence and security policies needed to be in place before a review of defence industrial policy could be undertaken. The pressing need for a policy framework for South Africa's arms industry was to emerge in other policy processes, and particularly through the work of the Cameron Commission.

---

[10] For a discussion of the ways in which the legacies of accommodation and compromise from the negotiation period have impacted on the policy processes of the post-apartheid era, see Ohlson, T., 'South Africa: from apartheid to multi-party democracy', *SIPRI Yearbook 1995: Armaments, Disarmament and International Security* (Oxford University Press: Oxford, 1995), pp. 139–45.

[11] Cameron Commission of Inquiry into Alleged Arms Transactions between Armscor and one Eli Wazan and other Related Matters, *First Report* (Cameron Commission: Johannesburg, June 1995); and Cameron Commission of Inquiry into Alleged Arms Transactions between Armscor and one Eli Wazan and other Related Matters, *Second Report* (Cameron Commission: Cape Town, Nov. 1995).

[12] Transitional Executive Council Sub-Council on Defence, National Policy for the Defence Industry (Sub-Council on Defence: Pretoria, Apr. 1994).

The White Paper on National Defence, approved by parliament in May 1996, did not include any major policy pronouncements on the future of the domestic arms industry. It affirmed the need for it for 'the maintenance, upgrading, and where necessary, the replacement of weapons and equipment' and stated that the government would encourage the industry 'to convert [its] production capability to civilian manufacture without losing the key technological capability needed for military production'.[13] Apart from these rather general statements, it stated that the government would prepare a separate White Paper on the arms industry in consultation with parliament, stakeholders and interest groups.[14] This suggestion for a separate White Paper echoed the recommendations of the second report of the Cameron Commission.[15]

Another highly significant consultative process, the Defence Review, was initiated in February 1996. It was concerned with long-range planning on matters such as doctrine, posture, force design, force levels, logistical support, armaments, equipment, human resources and funding, and was derived from the policy framework contained in the White Paper on National Defence.[16] While it was not specifically concerned with the future of the domestic arms industry, it dealt with matters relating to armaments procurement and therefore considered some policy issues relating to the arms industry in an indirect way. The Defence Review was completed, approved by the cabinet in June 1997 and by the parliament in August 1997.

Another significant development with regard to policy on the domestic arms industry was the formation in July 1994 of the South African Defence Industry Association (SADIA). The main functions of SADIA are to coordinate the activities of the domestic arms industry and to act as a mouthpiece for the industry in dealings with government and other interested parties.[17] These functions were previously undertaken by Armscor on behalf of the industry. SADIA has played a significant, and sometimes controversial, role in the various policy processes described above. During the course of the Cameron Commission public hearings it consciously distanced itself from the actions of Armscor, thus undermining the 'shared interests' which had existed in the arms industry during the apartheid era. SADIA has publicly criticized the lack of government policy on the arms industry, arguing that it has caused uncertainty within the industry and already led to the outflow of technology and know-how out of the country in the form of off-shore alliances and joint ventures.[18] The faster-than-expected phasing out of the GEIS has also drawn criticism from SADIA. Its Executive Director, Julius Kriel, has said that 'to compete overseas some government assistance is necessary'.[19]

[13] South African Department of Defence, *Defence in a Democracy: White Paper on National Defence* (Government Printer: Pretoria, 1996), p. 41.
[14] *White Paper on National Defence* (note 13), p. 41.
[15] Cameron Commission, *Second Report* (note 11).
[16] *White Paper on National Defence* (note 13), p. 41.
[17] 'What is SADIA?', *Engineering News*, 1 Mar. 1996, p. 23.
[18] Interview with Maj.-Gen. (ret.) Julius Kriel, *Engineering News*, 1 Mar. 1996, p. 23.
[19] 'Nearly 50 000 defence industry jobs lost: association', *Business Day*, 14 Jan. 1997.

Despite the absence of a clearly articulated government policy on the industry, many senior officials within the Ministry of Defence have made public pronouncements on the future of the arms industry and have highlighted the strategic, political, economic and psychological arguments in favour of maintaining it.[20] Joe Modise has stated that South Africa needs a capable arms industry for strategic reasons because of the 'instability around us' and the need to be able 'to protect against any instability that spills over' into South Africa from the region. He also argues that the industry can play a greater role in the region in order to promote a high degree of regional equipment standardization.[21] Pierre Steyn, the Secretary of Defence, has highlighted the strategic reasons for maintaining the domestic arms industry in order to 'meet the key technology needs of the SANDF and to function as a technology multiplier in the deployment of South Africa's future industrial strategy'.[22]

In contrast to the high-profile support for maintaining the arms industry from within the defence establishment, certain sections of civil society, including the churches, the trade unions and NGOs, have argued in favour of the further diversification and conversion of the domestic arms industry.[23] The debate about the future of the domestic arms industry is a healthy sign of the increasing transparency and accountability on defence matters in South Africa and reflects the increasing involvement of civil society in the formulation of policy on defence matters.

After long delays and amid growing frustration within the arms industry, in March 1997 the government initiated the preparation of a White Paper on the Defence Industry. It is being coordinated by Professor Kader Asmal, Chairman of the NCACC, and the Defence Secretariat. A working group consisting of representatives from Armscor, Denel, the arms industry, government departments (trade and industry, science and technology, and foreign affairs), the Council for Scientific and Industrial Research (CSIR), NGOs and universities has been established to guide the process of producing the White Paper. It is expected to be completed by the first half of 1998.

## IV. Arms trade policy and the arms industry

One area in which the government has been very proactive on behalf of the arms industry has been the promotion of arms exports. With the advent of the ANC-led government, it was assumed that a new morality would be applied to South Africa's arms trade. However, President Mandela has defied expectations by publicly defending the arms trade and by promoting arms sales on many of

[20] 'Modise highlights benefits of SA defence industry' (note 9).
[21] Interview with Defence Minister Joe Modise (note 8), p. 32.
[22] 'The South African security environment: a perspective on the defence policy formulation', Speech to the Conference on the Future of the South African Defence Industry, Armscor, SADIA and Idasa, Midrand, 28–30 Mar. 1996.
[23] Submissions by the Black Sash and the Anglican Church to the Cameron Commission Public Hearing, Cape Town, June 1995.

his trips overseas.[24] At the same time, however, both Mandela and Modise have stated their support for a responsible arms industry and for better oversight and control over arms exports.[25]

On 25 May 1994 the UN lifted the arms embargo against South Africa by revoking Security Council resolutions 418 of 1977 and 558 of 1984 concerning, respectively, the supply of defence equipment to and the purchase of arms from South Africa. With unseemly haste Armscor announced its intention to increase South Africa's share of the global arms market from 0.4 per cent in 1994 to 2 per cent, increasing the value of sales from 800 million rand to roughly 2.4 billion rand.[26] The decline in domestic defence expenditure, and particularly the procurement budget, created the incentive to maximize arms exports; at the same time, however, there has been a recognition on the part of the new government of the need to eradicate past arms trade practices which have proved detrimental to South Africa's newly acquired international standing.

The pursuit of international legitimacy and the desire to shed the reputation of being a pariah state have translated into practical measures designed to radically transform South Africa's arms trade policy by developing a more open and transparent approach to arms transfer decisions.[27] Nevertheless, many contradictions persist in South Africa's arms trade policies, reflecting the tensions between the need to offset the decline in domestic demand, the adoption of international norms and standards, and the pursuit of a non-aligned foreign policy.

In September 1994 it emerged that a consignment of South African arms supposedly destined for Lebanon had apparently been sold to Yemen, a prohibited destination for South African arms. As a result of the international outcry which followed, President Mandela appointed a commission of inquiry, the Cameron Commission, to investigate the details of the aborted arms deal with Lebanon and to comment on South Africa's existing arms trade policies and decision-making procedures.[28] The first report of the Cameron Commission, which was released in June 1995, was not directly concerned with policy on the arms industry. However, it was highly critical of Armscor and made a number of

[24] 'The world is an oyster for the SA arms industry', *The Star,* 14 Apr. 1996, p. 1. See also chapter 6, section V in this volume.

[25] 'Taming dogs of war', *The Sowetan*, 10 Sep. 1997, p. 1. For a discussion of these issues, see George, P., 'The impact of South Africa's arms sales policy on regional military expenditure, development and security', ed. Swedish Ministry of Foreign Affairs, *Säkerhet och utveckling i Afrika* [Security and development in Africa] (Utrikesdepartementet: Stockholm, Nov. 1995), appendix, pp. 237–96 [appendix in English]; and Singh, R. P. and Wezeman, P. D., 'South Africa's arms production and exports', *SIPRI Yearbook 1995* (note 10), appendix 14E, p. 575.

[26] Willett, S. and Batchelor, P., *To Trade or Not to Trade: The Costs and Benefits of South Africa's Arms Trade*, Working Paper no. 9 (Military Research Group: Johannesburg, 1994), p. 1.

[27] For a discussion of South Africa's arms trade and international norms, see Willett, S., 'Open arms for the prodigal son? The future of South Africa's arms trade policies', *African Defence Review*, no. 17 (July 1994).

[28] Cilliers, J., 'Towards a South African conventional arms trade policy', *African Security Review*, vol. 4, no. 4 (1995), pp. 3–20.

recommendations concerning the transformation of Armscor and the need for new and tighter controls over South African arms exports.[29]

In the light of revelations which emerged during the hearings of the Cameron Commission, in March 1995 the cabinet appointed a ministerial committee under the Defence Minister to investigate the issue of South Africa's conventional arms trade policy and to make recommendations to the government on policy and a code of conduct. The Modise Commission submitted its proposals in August 1995. As a result of the findings of the Cameron Commission and the proposals of the Modise Commission, the cabinet established new policy and procedures for arms control, which included four levels of control, the highest being the NCACC established in August 1995 under Minister Asmal.[30] It took over the arms control function which had been carried out by Armscor since the early 1980s.[31] It was also mandated to formulate policy on the arms industry, and thus took over from a cabinet committee on the arms industry established in August 1994 under the Minister of Defence.[32] The NCACC consists of the ministers for defence, trade and industry, foreign affairs, safety and security, general services, the arts, culture, and science and technology, and the deputy ministers of defence, foreign affairs and intelligence services, and safety and security and is directly answerable to the cabinet.

The second report of the Cameron Commission, published in March 1996, contained a number of recommendations with respect to arms trade policy- and decision-making processes. These included a proposal to classify countries according to their status as potential customers: those to whom arms should not be sold, those to whom arms could be sold and those about which the situation is unclear. A further recommendation was that parliament should be given the right to review and veto arms sales. Both these recommendations have been rejected by Asmal.[33] The second report also recorded the submissions to the public hearings and challenged many of the political, strategic and economic arguments in favour of South Africa's continued involvement in the international arms trade. It recommended the restructuring of the domestic arms industry and suggested that the 'future of the [defence] industry, and the question of conversion to civilian production should be the subject of a White Paper'.[34] The second report has been widely criticized by a number of commentators, who argue that implementation of its recommendations would 'severely restrict, complicate and probably end any significant defence exports'.[35]

[29] Cameron Commission, *First Report* (note 11), pp. 130–34.

[30] Cabinet Memorandum: National Conventional Arms Control Committee, Cabinet Office: Pretoria, 30 Aug. 1995. See also *White Paper on National Defence* (note 13), pp. 53–58.

[31] van Dyk, J., 'Conventional arms control: a new system for a new era', *SALVO*, vol. 1 (1996), p. 2. The fact that Armscor had responsibility for the marketing of arms and arms control was identified as a problem in the Draft National Policy for the Defence Industry submitted to the TEC Sub-Council on Defence in Apr. 1994.

[32] van Dyk (note 31).

[33] 'Inquiry lashes SA trade in arms of death', *The Star*, 26 Mar. 1996; and 'Arms deals to remain undisclosed', *Weekly Mail and Guardian*, 29 Mar. 1996.

[34] Cameron Commission, *Second Report* (note 11), p. vi.

[35] 'Findings of Cameron enquiry into arms trade pre-empted by govt action', *Business Day*, 23 Apr. 1996.

Despite the attempts of the Cameron and Modise commissions to recast South Africa as a responsible arms seller, the government appears to have flouted its own guidelines on a number of well-publicized occasions.[36] The NCACC's approval of small-arms transfers to Rwanda in September 1996 was a case in point.[37] A loss of faith in the NCACC has given rise to calls for the establishment of a special parliamentary arms committee to monitor cabinet arms trade decision making.[38]

Another scandal which aroused much public condemnation and has exposed the contradictions in South Africa's present arms trade policies and practices was the proposed sale to Syria of electronic tank sightings for Soviet-built T-72 tanks.[39] In this case Israel warned South Africa that approval of the deal could upset the delicate peace process in the Middle East, while the USA warned it that if it went ahead with arms sales to a country which the United States regards as a state sponsor of terrorism it would impose sanctions, notably by cutting $121 million-worth of development aid.[40] The domestic outcry was no less censorious. Opposition parties within South Africa used the issue to attack the ANC-dominated government, pointing out how transfers to Syria were damaging the country's international reputation and its stand on human rights. President Mandela responded to the international outcry by arguing that the South African Government had a right to decide on its own foreign policy priorities and that he was not prepared to bow to pressure from abroad.[41]

The ANC has been adamant about its right to pursue a non-aligned foreign policy. It maintains a strong sense of debt to those countries which gave it support during the apartheid days—hence the warmth of its relations with China, Cuba, Iran and certain Arab states. So far it has resisted international pressure to abandon its controversial policy of non-alignment. Its initial reaction to US criticisms of arms sales to Syria was to accuse the USA of seeking to bully and blackmail South Africa into submission, although it toned down its language in later exchanges.

The government risks jeopardizing its recently acquired international good-will if it persists with plans to sell arms to such controversial destinations, but having opted for an arms export maximization policy in order to keep the domestic arms industry afloat it faces a number of dilemmas. South African defence firms face considerable barriers in breaking into major world markets because of the overwhelming dominance of the major suppliers (France, Russia, the UK and the USA) and the cut-throat nature of the global arms market, especially since the end of the cold war. This has encouraged the industry to target more controversial arms markets, such as China, Indonesia and Syria, as well as African purchasers such as Rwanda, Sudan and Uganda. This in turn

[36] See, e.g., 'Arms for Syria?', *The Economist*, 18 Jan 1997; and 'South Africa: Damascus deal', *Oxford Analytica Daily Brief*, 22 Jan. 1997.
[37] 'Storm over SA approval of arms sales to Rwanda', *Sunday Independent*, 29 Sep. 1996, p. 1.
[38] *Oxford Analytica* (note 36).
[39] *Oxford Analytica* (note 36).
[40] 'Realpolitik tests Mandela's old friendships', *Financial Times*, 22 Jan. 1997.
[41] *Financial Times* (note 40).

**Table 7.1.** Military expenditure, 1989–95

|  | 1989 | 1990 | 1991 | 1992 | 1993 | 1994 | 1995 |
|---|---|---|---|---|---|---|---|
| Military expenditure (m. rand, constant 1990 prices) | 11 435 | 10 070 | 8 094 | 7 605 | 6 589 | 7 153 | 6 233 |
| % change on previous year | 5.7 | – 11.9 | – 19.6 | – 6.0 | – 13.4 | 8.6 | – 12.9 |
| Military exp. as share of GDP (%) | 4.1 | 3.6 | 2.9 | 2.8 | 2.4 | 2.5 | 2.2 |
| Military exp. as share of gov't exp. (%) | 13.0 | 12.4 | 9.8 | 8.4 | 6.8 | 8.1 | 6.8 |

*Sources:* South African Department of State Expenditure, *Printed Estimates of Expenditure* (South African Department of State Expenditure: Pretoria, various years); and South African Department of Finance, *Budget Review 1995* (South African Department of Finance: Pretoria, various years).

undermines the government's aspirations to take the moral high ground on human rights and ethical issues in its external relations. Thus, although several institutional and normative reforms have been instituted, so far the government appears to have failed to secure the right balance between restraint, responsibility and arms export maximization.

## V. Defence budget trends since 1994

Given the government's commitment to funding the RDP and reducing the budget deficit, the defence budget continued to be cut after April 1994. The Minister of Defence, speaking in the 1995 defence budget vote, stated that 'we are living in a situation of relative peace and stability . . . the country faces no specific threat in the near future . . . we wish to cut defence spending to the bone'.[42] Despite acknowledging the need for cuts, the Department of Defence has on a number of occasions cautioned against further defence cuts, arguing that the current allocation does not adequately provide for the maintenance and replacement of equipment.[43] Despite its warnings, the defence budget has continued to decline in real terms from 2.4 per cent of GDP in 1993 to just over 2 per cent in 1995 (see table 7.1).

It is evident from the table that the defence budget increased in real terms in 1994 before declining again in 1995. The increase in 1994 was largely a result of the significant costs associated with the integration process, which started immediately after the elections in April 1994 and was expected to be completed by the end of 1999.[44] As a result of the integration and rationalization process, the share of personnel and operating expenditure increased quite significantly in 1994, largely at the expense of procurement. The share of current expend-

---

[42] Defence Budget Speech, National Assembly, 24 May 1995.

[43] 'Defence budget suffers a severe blow', *Weekly Mail and Guardian*, 19 Apr. 1996.

[44] The costs of integration and rationalization were estimated at 2.2 billion rand in 1995 prices. Briefing on Defence Budget to Joint Standing Portfolio Committee on Defence, 4 Apr. 1995.

**Table 7.2.** The distribution of military expenditure by function, 1989–95
Figures are percentages.

|  | 1989 | 1990 | 1991 | 1992 | 1993 | 1994 | 1995 |
|---|---|---|---|---|---|---|---|
| Personnel | 18.9 | 21.2 | 27.7 | 27.9 | 31.6 | 38.6 | 33.8 |
| Operating | 22.6 | 21.7 | 26.9 | 26.9 | 28.3 | 33.5 | 32.8 |
| Procurement[a] | 58.5 | 57.1 | 45.4 | 45.2 | 40.1 | 27.9 | 33.4 |
| Total | 100.0 | 100.0 | 100.0 | 100.0 | 100.0 | 100.0 | 100.0 |
| R&D[b] | 8.6 | 7.9 | 7.2 | 6.1 | 5.2 | 4.8 | 5.4 |

[a] All procurement expenditure is funded from the Special Defence Account.

[b] R&D expenditure is funded from operating (General Support Programme) and procurement (Special Defence Account) expenditure.

*Source:* South African Department of State Expenditure, *Printed Estimates of Expenditure* (South African Department of State Expenditure: Pretoria, various years).

iture (personnel and operating costs) increased from 60 per cent in 1993 to over 66 per cent in 1995, while the share of capital expenditure (procurement of armaments) declined from 40.1 per cent in 1993 to 33.4 per cent in 1995 (see table 7.2). Neither the value nor the share of military R&D fell much after 1993, despite the dramatic real cuts in procurement expenditure (see table 7.3).

Despite the cuts in procurement expenditure, the SANDF has continued to acquire new equipment since April 1994. Major procurement projects for the army include the Rooikat armoured car, the Mamba light APC, upgraded and retrofitted GV6 howitzers and the Zumlac self-propelled anti-aircraft gun.[45] The air force's major procurement projects include 60 Pilatus PC-7 Mk II training aircraft from Switzerland, a project which started in 1994 and will be completed during 1996–99,[46] 51 medium transport helicopters (Oryx) and the upgrading of 38 Cheetah fighter aircraft (in conjunction with IAI/Elta in Israel).[47] Programmes are already under way to develop short- and medium-range air-to-air missiles, and in April 1996 it was announced that the air force would purchase 12 Rooivalk attack helicopters, down from the original requirement of 16.[48] The next major procurement project, for which funding has yet to be secured, is the purchase of (foreign) jet trainers to replace the existing Impala Mk II.[49] The navy's major procurement projects include four new corvettes (the hulls will be purchased from abroad and fitted with South African combat systems) and the upgrading of the three existing Daphne Class submarines.[50]

[45] 'Post-embargo SAAF rebuilds', *Jane's Defence Weekly*, 18 Feb. 1995, p. 15.
[46] *Armscor Annual Report 1995/96* (note 2), p. 10.
[47] 'SAAF pointed to rule on multi-role Cheetahs', *Jane's Defence Weekly*, 10 Dec. 1994, p. 6.
[48] 'Rooivalk flies into further discord as its true costs become apparent', *Sunday Independent*, 28 Apr. 1996, p. 6.
[49] 'Aerospace companies target SA', *The Star*, 25 Apr. 1996.
[50] 'SA narrows corvette shortlist', *Jane's Defence Weekly*, 14 Jan. 1995, p. 12.

**Table 7.3.** Military research and development expenditure, 1989–95

|  | 1989 | 1990 | 1991 | 1992 | 1993 | 1994 | 1995 |
|---|---|---|---|---|---|---|---|
| Technology development[a] | 180 | 163 | 340 | 321 | 210 | 188 | 195 |
| (m. rand, constant 1990 prices) |  |  |  |  |  |  |  |
| as share of total (%) | 18.2 | 20.6 | 58.7 | 68.8 | 61.2 | 55.0 | 57.2 |
| Procurement projects | 806 | 630 | 240 | 146 | 133 | 154 | 147 |
| (m. rand, constant 1990 prices) |  |  |  |  |  |  |  |
| as share of total (%) | 81.8 | 79.4 | 41.3 | 31.2 | 38.8 | 45.0 | 42.8 |
| Total military R&D[b] | 985 | 793 | 580 | 467 | 343 | 342 | 342 |
| (m. rand, constant 1990 prices) |  |  |  |  |  |  |  |
| % change on previous year |  | – 19.5 | – 26.9 | – 19.5 | – 26.8 | 0.0 | – 0.2 |
| Military expenditure | 11 435 | 10 070 | 8 094 | 7 605 | 6 589 | 7 153 | 6 233 |
| (m. rand, constant 1990 prices) |  |  |  |  |  |  |  |
| R&D as share of military exp. (%) | 8.6 | 7.9 | 7.2 | 6.1 | 5.2 | 4.8 | 5.5 |

[a] Includes the Industry and Technology Survival Plan (ITSP) and the Technology Development Programme (TDP).

[b] R&D expenditure is funded from operating (General Support Programme) and procurement (Special Defence Account) expenditure.

*Sources:* South African Department of State Expenditure, *Printed Estimates of Expenditure* (South African Department of State Expenditure: Pretoria, various years); and private communications with Armscor officials.

The bulk of the current procurement projects are geared to maintaining and upgrading the defence forces' existing equipment, filling equipment gaps and maintaining the existing, largely offensive, across-the-board capabilities. However, they are also being used to ensure the survival of the domestic arms industry and to support the industry's export drive.[51] The air force has consumed the lion's share of the procurement budget in the past 10 years and is likely to continue to do so for the next 10 years. The navy's share of procurement expenditure, which was relatively insignificant in the past 10 years, will increase substantially over the next 10 years if the corvette project is approved (see table 7.4).

The lifting of the UN arms embargo has already lowered the prices of imported components and spares, which has meant that the SANDF can afford to maintain and keep in service some equipment which is relatively old and nearing retirement, such as the Daphne Class submarines and Dakota maritime patrol aircraft.[52] The absence of the UN embargo will also in all likelihood lead to a larger import share in future procurement expenditure, especially for aircraft and ships—the two main areas in which the domestic arms industry has not acquired a high degree of self-sufficiency.

---

[51] The air force's purchase of the Rooivalk is intended to assist with export sales of the helicopter. 'Denel expects a cornucopia from Rooivalk sales', *Sunday Independent*, 12 May 1996.

[52] At the height of the UN arms embargo South Africa paid a premium on imported components and spares which was sometimes as much as 10%. Private communication with Armscor official, Apr. 1994.

**Table 7.4.** Major current procurement projects, 1995

Figures are in m. rand in current prices. Figures in italics are in percentages.

| Project | Total (1995)[a] | Total[b] |
|---|---|---|
| *Army* | | |
| 242 Rooikat IFVs | 1 738 | 3 619 |
| 1094 Mamba APCs | 208 | 1 032 |
| Tactical communications system | 442 | 870 |
| Upgrading G6 guns | 663 | 798 |
| Anti-aircraft system | 464 | 1 202 |
| 110 Olifant MBTs | 773 | 1 455 |
| Electronic warfare system | 89 | 547 |
| Other | 1 794 | 4 141 |
| **Army total** | **6 171** | **13 664** |
| *Air force* | | |
| 38 Cheetah C fighters | 6 074 | 6 548 |
| 60 Pilatus trainers | 662 | 890 |
| Medium-range air-to-air missile | 1 375 | 1 664 |
| 51 Oryx helicopters | 2 318 | 2 340 |
| Short-range air-to-air missile | 401 | 531 |
| Rooivalk attack helicopter[c] | 167 | 1 325 |
| Upgrade static radar | 517 | 555 |
| Radio communications for aircraft | 84 | 629 |
| Other | 3 671 | 19 763 |
| **Air force total** | **15 269** | **34 245** |
| *Navy* | | |
| 4 corvettes[d] | 219 | 1 736 |
| Upgrade submarines | 434 | 493 |
| Integrated communications for ships | 95 | 108 |
| Other | 228 | 6 035 |
| **Navy total** | **976** | **8 372** |
| SANDF operations total | 141 | 243 |
| **Total project expenditure** | **22 557** | **56 254** |
| Army as share of total (%) | *27.4* | *24.2* |
| Air force as share of total (%) | *67.7* | *60.8* |
| Navy as share of total (%) | *4.3* | *14.8* |
| Operations Division as share of total (%) | *0.6* | *0.2* |

*Notes:* IFV = infantry fighting vehicle; APC = armoured personnel carrier; MBT = main battle tank.

[a] Total expenditure on project up to and including 1995.

[b] Total expenditure on project up to and including 2004 (estimate).

[c] The figures in both columns are for expenditure on the engineering development model only and do not include expenditure for procurement of the helicopter for the air force (876 million rand).

[d] Project postponed until completion of the Defence Review.

*Sources:* Private communications with Armscor officials and SANDF.

The completion of the SANDF's demobilization and rationalization, expected by 1999, will result in a much smaller defence force and the possibility of an end to the SANDF's deployment internally in support of the police. It will also lift pressure on the equipment budget in the medium term and may result in an increase in the value and share of procurement expenditure after 1999.

The cuts in defence expenditure since April 1994 and the resulting heated public debate about the allocation of state resources in the post-apartheid era have been covered in the previous chapter. In addition to the internal political struggle over military expenditure and the future of the arms industry, there has been some external pressure on the government to keep its defence expenditure down to acceptable international levels. Multilateral organizations such as the World Bank and the International Monetary Fund (IMF) have informally adopted 2 per cent of GDP as an acceptable benchmark for military expenditure in developing countries. Both have linked high military expenditure to high levels of debt and the failure of development. Although South Africa has been fortunate in avoiding the necessity of an IMF structural adjustment programme, the World Bank is a major donor in South Africa and is thus able to apply pressure on the South African Government to keep military expenditure around the 2 per cent benchmark.[53]

# VI. Defence industrial adjustment since 1994

As a result of the further defence cuts since 1994, and in the absence of any clear policy direction from government, the domestic arms industry has been forced to continue with the process of restructuring and down-sizing which started in the early 1990s. Both public- and private-sector defence firms have continued to pursue various defensive (scaling down) and offensive (expansion) adjustment strategies in order to survive and to adjust to the declining domestic defence market. The government's 'hands off' approach to defence industrial adjustment has meant that adjustment strategies have been unfocused and largely driven by market forces.

## Adjustment strategies in the public-sector arms industry

### Armscor

Despite its history of covert and illegal activities during the apartheid era and the revelations of the Cameron Commission, Armscor has managed to survive in the post-apartheid era as the state's armaments procurement organization. It has also managed to retain its status as a statutory state corporation (like the Atomic Energy Corporation) rather than being absorbed into the new Department of Defence or the Defence Secretariat. It is, however, part of the Ministry

---

[53] Solomon, H. and Cilliers, J., *People, Poverty and Peace*, IDP Monograph Series no. 4 (Institute for Defence Policy: Halfway House, May 1996), p. 18.

of Defence and its Board is still accountable to the Minister of Defence rather than to the head of the Department of Defence (the Chief of the SANDF) or to the departmental accounting officer, the Defence Secretary.

Armscor has attempted to implement a number of changes in order to retain its central role in the domestic arms industry and thereby its influence within the state.[54] It has declared its support for the RDP and seconded staff to the RDP Management Office to assist with project management.[55] This is a highly significant example of its attempts to redefine itself and to embrace the government's agenda on development and redistribution. Some internal organizational changes have been a direct result of the recommendations of the first report of the Cameron Commission, and some have been undertaken in order to project a more accountable and representative public image.[56] Most important, the function of issuing arms export permits, which was previously vested in Armscor, has been transferred to the Defence Secretariat and the NCACC. Since the establishment of the NCACC Armscor's functions have thus been narrowed to include procurement (which includes technology development, system development and logistic support), marketing assistance, SANDF surplus stock sales, and managing offset arrangements.[57]

Armscor has continued to support the adjustment strategies of public- and private-sector defence firms through its own, largely offensive, adjustment strategies. Most of these have been directed towards restructuring Armscor and ensuring the survival of the domestic arms industry, which is the reason for Armscor's corporate existence.

As one offensive strategy, Armscor intensified its international marketing campaign after the lifting of the UN arms embargo, and during the next few years opened additional overseas offices to support the domestic arms industry's export drive.[58] It coordinated and assisted participation in international defence exhibitions (for instance, LIMA '95 in Malaysia) and helped to facilitate international joint ventures and alliances between South African and foreign defence firms.[59] It has actively pursued the sale of surplus SANDF weapons in order to increase the amount of money available on the Special Defence Account for purchases of new weapons and equipment.[60] Its role in

[54] 'Armscor searches for its particular niche in new set-up', *Business Day*, 3 Nov. 1995.

[55] Both Armscor and Denel have allocated resources to the RDP. *Armscor Annual Report 1994/95* and *1995/96* (note 2); and *Denel Annual Report 1994/95* and *1995/96*.

[56] T. de Waal, Managing Director of Armscor, resigned shortly after the publication of the first report of the Cameron Commission and was replaced by an acting Managing Director. The Board of Armscor was completely restructured in late 1995 as a result of the recommendations of the Commission and a new Chairman, R. Haywood, was appointed to replace I. Moolman, who had served as Chairman since the early 1990s. 'Experts join Armscor', *The Star*, 22 Sep. 1995.

[57] 'Armscor searches for its particular niche in new set-up' (note 54).

[58] Armscor's foreign offices are now in China, France, Israel, Malaysia, Russia, Switzerland, the United Arab Emirates and the USA. *Armscor Annual Report 1994/95* (note 55). See also chapter 5, note 20 in this volume.

[59] Armscor facilitated a protocol on military technology cooperation between South Africa and Russia which was signed during 1995. *Armscor Annual Report 1995/96* (note 2), p. 28.

[60] The Ministry of Finance allows the money earned from the sale of surplus weapons to be used in the Special Defence Account for purchases of new equipment and armaments. Sales of SANDF surplus

selling surplus weapons remains controversial, and in late 1995 the Office for Serious Economic Offences started investigating alleged irregularities in the commissions paid in the sale of 12 ex-SANDF Puma helicopters to Romania.[61]

South Africa's share of the international arms market was estimated at less than 1 per cent in 1995. Despite increases in the value of its arms exports in the 1990s, this share has not increased significantly since the late 1980s.[62]

Armscor has increased the counter-trade requirement (up to 60 per cent) on contracts with foreign suppliers since April 1994. This has been done to save foreign exchange and to support the domestic manufacturing economy, particularly the arms industry. It has stated that it aims to increase the counter-trade requirement to 70 per cent in 1996 and 100 per cent in 1998.[63]

Armscor has continued to use the defence budget, particularly the Special Defence Account, and some of its own resources to support R&D efforts in public- and private-sector defence firms and thereby maintain certain critical defence technologies and capabilities within the domestic arms industry. Since April 1994 it has also acquired a number of private-sector companies, such as Protechnik Laboratories (Pty), a R&D company specializing in chemical defence, and Hazmat Protective Systems (Pty) Ltd, which manufactures respiratory equipment for use in defensive chemical and biological warfare, in order to maintain certain strategic industrial and technological capabilities.[64]

The corporation has pursued a more transparent and competitive procurement policy in order to achieve better value for money. The establishment of the Accreditation of Defence Suppliers and the *Acquisition Bulletin* have contributed to a more transparent and competitive procurement process.[65] Even so, Armscor and the SANDF have been severely criticized for the lack of transparency and consultation with regard to both the corvette project and the Rooivalk attack helicopter project.[66]

*Denel*

The new government initially resisted the idea of privatizing Denel, given the nature of its business and the opposition to privatization of state assets from organized labour (such as the Congress of South African Trade Unions, COSATU). However, in late 1995 the government announced that it was investigating the restructuring of state assets (that is, privatization), including Denel's.[67] By mid-1997 no decision on privatization of Denel had been taken.

weapons by Armscor amounted to 66 million rand in 1994 and 64 million rand in 1995. *Armscor Annual Report 1995/96* (note 2), p. 30.
[61] 'Probe into Armscor chopper sale', *The Star*, 2 Oct. 1995.
[62] 'Armscor chief predicts defence export boost', *The Star*, 12 Dec. 1995. See also 'SA supplied less than 1 percent of world arms', *The Citizen*, 19 Mar. 1996.
[63] 'Armscor clients to face stricter countertrade criteria by 1998', *Business Day*, 22 Dec. 1995.
[64] *Armscor Annual Report 1995/96* (note 2), pp. 19–20.
[65] *Armscor Annual Report 1995/96* (note 2), p. 27.
[66] *Sunday Independent*, 28 Apr. 1996, p. 6.
[67] 'Privatisation: a first step on the high road', *Financial Mail,* 8 Sep. 1995.

**Table 7.5.** Denel exports, 1992–95

|  | 1992 | 1993 | 1994 | 1995 |
|---|---|---|---|---|
| Arms exports (m. rand, constant 1990 prices) | 353 | 406 | 439 | 492 |
| % change |  | 15.0 | 7.9 | 12.2 |
| Civilian exports (m. rand, constant 1990 prices) | 23 | 48 | 80 | 110 |
| % change |  | 111.2 | 66.5 | 37.3 |
| Total exports (m. rand, constant 1990 prices) | 376 | 454 | 519 | 602 |
| as share of turnover (%) | 17 | 23 | 26 | 30 |
| Arms as share of total exports (%) | 93.8 | 89.4 | 84.5 | 80.7 |
| Civilian exports as share of total exports (%) | 6.2 | 10.6 | 15.5 | 18.3 |

*Source: Denel Annual Report*, various years.

The downward trend in defence expenditure has had a negative impact on Denel and the company has continued to pursue a variety of defensive and offensive adjustment strategies. In terms of defensive strategies, it continued to lay off employees during 1994 and 1995. By 1995 employment in the company had stabilized at 14 150, down from over 15 500 in 1992. Denel also continued to rationalize and consolidate its defence and non-defence activities after 1994, and this resulted in the closure of certain facilities (Houwteq being one). The company found it particularly expensive to maintain some of the testing facilities which it inherited from Armscor (such as Gerotek and the Overberg Test Range) and which did not provide a viable return.

All Denel's groups and their respective divisions and business units continued to pursue diversification strategies after 1994, particularly the acquisition of non-defence products or firms, mergers and joint ventures with civilian firms, and the development of civilian products derived from existing defence technologies and products. The strategy of acquisitions and mergers with civilian firms is regarded as the most successful diversification strategy but has proved problematic since Denel is still 100 per cent owned by the state. Thus joint ventures and alliances are seen as the most appropriate diversification strategies at the moment. The strategy of conversion has been largely abandoned, given the significant difficulties and costs involved in converting facilities to civilian use and the expensive failure of Houwteq's conversion effort.[68]

In terms of offensive strategies, Denel has continued to pursue export markets since April 1994, particularly in the light of the lifting of the UN arms embargo. The value and share of its arms export business witnessed significant increases in 1994 and 1995 (see table 7.5): defence exports rose from 16 per cent of turnover in 1992 to 24 per cent in 1995 (see table 7.7).

The declining domestic defence market and the increasingly competitive international arms market have encouraged Denel to enter into a number of joint ventures and strategic alliances with foreign defence firms, but here again the fact that it is still 100 per cent state-owned has prevented it from pursuing

---

[68] Private communication between the authors and Denel officials, Feb. 1997.

**Table 7.6.** Denel: international joint ventures, 1996

| Division | Foreign company | Country | Product/Project |
|----------|-----------------|---------|-----------------|
| Atlas Aviation | Eurocopter | France/Germany | Helicopters |
| Atlas Aviation | IAI/Elta | Israel | Cheetah upgrades |
| Atlas Aviation | Sextant Avionique | France | Rooivalk avionics |
| Atlas Aviation | Marshall Aerospace | UK | C-130 upgrades |
| Atlas Aviation | Airod | Malaysia | Rooivalk/Oryx |
| Atlas Aviation | British Aerospace | UK | Electronics |
| LIW | GEC-Marconi | UK | Anti-aircraft turret on G6 chassis |
| LIW | GEC-Marconi | UK | eGlas 35-mm gun |
| LIW/Kentron | Bumar Labedy | Poland | T-72 tank |
| Eloptro | Malaysia Optronic | Malaysia | Night vision equipment |
| Kentron | Oerlikon Contraves | Switzerland | Missiles |
| Kentron | GEC-Marconi | UK | Missile sub-system |
| Kentron | Pilkingtons | UK | Helmet-mounted sights |
| Kentron | Siemens Plessey | UK | Air defence radar |
| Mechem | Royal Ordnance | UK | Mine clearance |
| Mechem | Alvis | UK | Vehicle design |

*Sources:* Private communications with Denel officials; *Denel Annual Report*, various years; South African Defence Industry Association (SADIA); and Singh, R. P. and Wezeman, P. D., 'South Africa's arms production and exports', *SIPRI Yearbook 1995: Armaments, Disarmament and International Security* (Oxford University Press: Oxford, 1995), appendix 14E.

acquisitions of or mergers with foreign companies. The vast majority of these international arrangements have been with British defence companies and have been concentrated on the Rooivalk attack helicopter and the potential replacement of the existing Impala Mk II jet trainers (see table 7.6).[69]

Another offensive adjustment strategy has been an attempt to monopolize certain sectors of the domestic defence market by integrating vertically and by subcontracting or outsourcing significantly less of its defence work than in the past. However, despite these efforts to maintain or increase its share of the domestic defence market, the value and share of Denel's domestic defence business have continued to decline in real terms since 1992 as a result of the cuts in defence expenditure.

The results of these adjustment strategies are reflected in the changing share of Denel's business. The share of domestic defence in turnover has fallen from 63 per cent in 1992 to 45 per cent in 1995; the share of civilian business (both domestic and export) has risen from 21 per cent to 31 per cent (see table 7.7).

Despite this success in increasing the value and share of its civilian business, Denel's financial performance in terms of profitability (operating margin) and asset management (return on assets) are still poor compared to those of the three major private-sector defence groups, Altech, Grintek and Reunert (see table 7.9). Profitability was mixed during 1994 and 1995. The comparatively

[69] 'Aerospace companies target SA', *The Star*, 25 Apr. 1996.

**Table 7.7.** Denel: composition of turnover, 1992–95

|  | 1992 | 1993 | 1994 | 1995 |
|---|---|---|---|---|
| Turnover (m. rand, constant 1990 prices) | 2 208 | 2 017 | 2 001 | 2 017 |
| % change |  | –8.7 | –0.8 | 0.8 |
| Domestic arms sales (m. rand, constant 1990 prices) | 1 391 | 1 069 | 932 | 907 |
| as share of turnover (%) | 63 | 53 | 47 | 45 |
| Arms exports (m. rand, constant 1990 prices) | 353 | 406 | 439 | 492 |
| as share of turnover (%) | 16 | 20 | 23 | 24 |
| Domestic civilian sales (m. rand, constant 1990 prices) | 441 | 493 | 491 | 508 |
| as share of turnover (%) | 20 | 24 | 25 | 25 |
| Civilian exports (m. rand, constant 1990 prices) | 23 | 49 | 80 | 110 |
| as share of turnover (%) | 1 | 3 | 5 | 6 |

*Source: Denel Annual Report*, various years.

poor return on assets is related to the large number of highly capital-intensive facilities inherited from Armscor which are not commercially viable. Turnover continued to fall in 1994, with a slight improvement in 1995.

In the absence of any clear policy direction from government, Denel will continue to pursue new defence and civilian markets and attempt to reduce its dependence on the local defence market. Its success in civilian markets will probably be limited to those divisions such as Informatics which use dual-use or generic technologies.[70] Some such as LIW which are prime contractors and heavily dependent on defence business will continue to find it difficult to diversify out of defence.

## Adjustment strategies in the private-sector arms industry

The private-sector arms industry, like the public-sector, has continued to pursue the same adjustment strategies it pursued during the transition period. Defensive adjustment strategies have included laying off workers, down-sizing and rationalizing defence operations, selling off assets and exit from the defence market. Most firms have also continued to pursue diversification, mainly through mergers with and acquisitions of civilian companies, and the purchase of civilian products or product lines. Fewer firms have pursued diversification through the development of civilian products based on defence products and technology, and almost none have pursued conversion at the plant or firm level.

Offensive adjustment strategies have included increasing exports, international joint ventures and alliances with foreign defence firms, concentration and monopolization of local defence markets. The increasing internationalization of South Africa's arms industry after April 1994 corresponded with the

[70] Denel Informatics reduced the share of defence business in its turnover from 63% in 1992 to 48% in 1994. 'Denel Informatics: muscling in on the private sector', *Financial Mail*, 25 Aug. 1995.

**Table 7.8.** Denel: financial performance, 1992–95

|  | 1992 | 1993 | 1994 | 1995 |
|---|---|---|---|---|
| Turnover (m. rand, constant 1990 prices) | 2 208 | 2 017 | 2 001 | 2 017 |
| % change on previous year |  | – 8.7 | – 0.8 | 0.8 |
| Operating profit (m. rand, constant 1990 prices) | 132 | 147 | 126 | 113 |
| Operating margin$^a$ | 6.0 | 7.3 | 6.5 | 5.6 |
| Net assets (m. rand, constant 1990 prices) | 2 835 | 2 711 | 2 150 | 2 002 |
| Return on assets$^b$ | 4.7 | 5.4 | 6.0 | 5.7 |
| Total employment (no. of staff) | 15 572 | 13 895 | 13 826 | 14 150 |
| Cost of sales$^c$ | 53.7 | 52.5 | 53.3 | 49.6 |
| Capital : output ratio (rand)$^d$ | 1.9 | 1.8 | 1.3 | 1.2 |
| Output : labour ratio (th. rand)$^e$ | 76.0 | 76.1 | 77.1 | 70.7 |
| Capital : labour ratio (th. rand)$^f$ | 182.0 | 195.0 | 155.5 | 141.4 |

$^a$ Operating profit/turnover.
$^b$ Operating profit/net assets (capital employed).
$^c$ Value added/turnover.
$^d$ Net assets/value added.
$^e$ Value added/total employment.
$^f$ Net assets/total employment.
*Source: Denel Annual Report*, various years.

**Table 7.9.** Denel and private-sector defence firms: financial performance, 1995

|  | Denel | Reunert | Grintek | Altech |
|---|---|---|---|---|
| Turnover (m. rand, current prices) | 3 401 | 4 742 | 2 044 | 1 593 |
| Operating margin$^a$ | 5.6 | 7.7 | 6.3 | 8.4 |
| Return on assets$^b$ | 5.7 | 29.1 | 32.6 | 16.5 |
| Total employment (no. of staff) | 14 150 | 15 938 | 3 009 | 3 494 |
| Cost of sales$^c$ | 49.6 | 35.0 | 21.5 | 36.7 |
| Capital : output ratio (rand)$^d$ | 2.0 | 0.7 | 0.9 | 1.3 |
| Output : labour ratio (th. rand, current prices)$^e$ | 119 | 104 | 146 | 131 |
| Capital : labour ratio (th. rand, current prices)$^f$ | 238 | 78 | 131 | 181 |

$^a$ Operating profit/turnover.
$^b$ Operating profit/net assets (capital employed).
$^c$ Value added/turnover.
$^d$ Net assets/value added.
$^e$ Value added/total employment.
$^f$ Net assets/total employment.
*Source:* Company annual reports, various years.

globalization of arms production, which started during the mid-1980s and has become a significant feature of the post-cold war international arms industry.[71]

[71] Bitzinger, R., *The Globalisation of Arms Production: Defence Markets in Transition* (Defence Budget Project: Washington, DC, 1993), pp. 5–15.

**Table 7.10.** Arms sales and exports, 1992–95

|  | 1992 | 1993 | 1994 | 1995 |
|---|---|---|---|---|
| Total South African arms sales[a]<br>(m. rand, constant 1990 prices) | 3 079 | 3 250 | 2 643 | 2 422 |
| % change on previous year | – 24.7 | 5.5 | – 18.6 | – 8.3 |
| Denel total arms sales<br>(m. rand, constant 1990 prices) | 1 745 | 1 476 | 1 371 | 1 400 |
| as share of total arms sales (%) | 56.7 | 45.4 | 51.9 | 57.8 |
| Private-sector total arms sales<br>(m. rand, constant 1990 prices) | 1 334 | 1 774 | 1 272 | 1 022 |
| as share of total arms sales (%) | 43.3 | 54.6 | 48.1 | 42.2 |
| Total South African arms exports<br>(m. rand, constant 1990 prices) | 382 | 625 | 550 | 613 |
| % change on previous year | – 45.3 | 63.5 | – 12.0 | 11.3 |
| Arms exports as % of total sales | 12.4 | 19.2 | 20.8 | 25.3 |
| Denel arms exports<br>(m. rand, constant 1990 prices) | 353 | 406 | 439 | 492 |
| as share of total arms exports (%) | 92.4 | 65.0 | 79.7 | 80.3 |
| Private-sector arms exports<br>(m. rand, constant 1990 prices) | 29 | 219 | 111 | 120 |
| as share of total arms exports (%) | 17.6 | 35.0 | 20.3 | 19.7 |
| Total South African arms sales as share of<br>manufacturing output (%) | 5.3 | 5.7 | 4.5 | 3.9 |
| Total SA arms sales as share of GDP (%) | 1.2 | 1.2 | 1.0 | 0.8 |

[a] Domestic procurement plus exports.

*Sources: Armscor Annual Report*, various years; *Denel Annual Report*, various years; and South African Defence Industry Association, *Some Statistics of the SADIA Members of the Defence Industry* (SADIA: Johannesburg, Jan. 1996).

Some private-sector firms, particularly the larger defence groups such as Reunert, have also attempted to monopolize and increase their share of certain sectors of the declining domestic defence market by reducing the amount of work that they subcontract to other firms, by purchasing smaller defence firms or by concentrating on certain market niches.

Most private-sector defence firms pursued arms export markets quite aggressively after April 1994, particularly with the lifting of the UN arms embargo (see table 7.10).

The private sector witnessed real declines in the value of its arms exports in 1994 after a peak in 1993, despite concerted efforts to market South African arms overseas. The extreme fluctuations in the value of private-sector arms exports between 1992 and 1995 were also a function of the highly competitive and volatile nature of the international arms market and the fact that arms export contracts (and payment) often take between three and five years to conclude. The total defence exports of Altech, Grintek and Reunert saw increases in 1992 and 1993, with a sharp decline in 1994 before a slight recovery in 1995.

**Table 7.11.** Private-sector international joint ventures, 1996

| Company | Foreign company | Country | Product |
|---|---|---|---|
| Reumech | ANI Corporation | Australia | Infantry vehicles |
| OMC (Reumech) | Vickers | UK | APCs |
| Sandock (Reumech) | Alvis | UK | Mine-protected vehicles |
| Reutech | GEC Marconi | UK | Radar |
| Reutech | Royal Ordnance | UK | Electronics |
| Reutech | Daimler Benz Aerospace | Germany | Naval radar |
| UEC Projects (Altech) | GEC Marconi | UK | Naval systems |
| UEC Projects (Altech) | Mafra Wira | Malaysia | Naval systems |
| Teklogic (Altech) | FATS | UK | Simulation |
| Grinaker Avitronics | Vinten | UK | Reconnaissance pods |
| Grinel | British Aerospace | UK | Radio products |
| Aerosud | Westland | UK | Helicopters |
| Aerosud | Marvol Group | Russia | Aircraft engines |
| ATE | British Aerospace | UK | Electronics |
| ATE | Airod | Malaysia | Avionics |
| ATE | Integrated Technologies and Systems | Malaysia | Electronic warfare |
| TFM | Westrac Equipment | Australia | Infantry vehicles |

*Note:* APC = armoured personnel carrier.

*Sources:* South African Defence Industry Association, *Some Statistics of the SADIA Members of the Defence Industry* (SADIA: Johannesburg, Jan. 1996); *Engineering News*, 1 Mar.1996, p. 27; and Singh, R. P. and Wezeman, P. D., 'South Africa's arms production and exports', *SIPRI Yearbook 1995: Armaments, Disarmament and International Security* (Oxford University Press: Oxford, 1995), appendix 14E.

These fluctuations were reflected at the level of individual companies. Reutech's defence exports as a share of total sales increased from 5 per cent in the early 1990s to 20 per cent in 1995,[72] while Grintek's rose from around 10 per cent of total sales in the early 1990s to nearly 16 per cent in 1994 before declining to 6 per cent in 1995.[73] Other private-sector firms, excluding the three major defence groups, witnessed real and sustained increases in the value of their arms exports between 1992 and 1995, although from a very low base.

Most private-sector defence firms pursued international joint ventures and/or alliances with foreign defence firms in an attempt to increase their export sales and to have a greater chance of winning foreign procurement contracts.[74] In 1996 the members of SADIA had a total of 93 joint ventures with firms in over 20 countries, the majority with companies in the UK (26), France (12), the USA (9), Germany (9) and Malaysia (9) (see table 7.11). Of these 29 per cent involved civilian technology, 63 per cent defence technology and the remaining

---

[72] *Engineering News*, 1 Mar. 1996, p. 27.
[73] Private communication with A. Holloway, Grinaker Avitronics, 13 May 1996.
[74] Singh and Wezeman (note 25), p. 587.

**Table 7.12.** The South African arms trade, 1989–95

|  | 1989 | 1990 | 1991 | 1992 | 1993 | 1994 | 1995 |
|---|---|---|---|---|---|---|---|
| Arms imports | 2 691 | 1 982 | 847 | 618 | 644 | 421 | 446 |
| (m. rand, constant 1990 prices) | | | | | | | |
| as % of manufactured exports | 6.0 | 5.0 | 2.1 | 1.6 | 1.5 | 0.8 | 0.7 |
| as % of total imports | 4.5 | 3.7 | 1.5 | 1.3 | 1.3 | 0.7 | 0.6 |
| Arms exports | 236 | 163 | 700 | 382 | 625 | 550 | 613 |
| (m. rand, constant 1990 prices) | | | | | | | |
| as % of manufactured exports | 0.9 | 0.7 | 3.0 | 1.9 | 3.2 | 2.5 | 2.2 |
| as % of total exports | 0.3 | 0.2 | 1.1 | 0.7 | 1.1 | 0.9 | 1.0 |
| Arms trade balance[a] | −2 455 | −1 819 | −147 | −236 | −19 | 129 | 167 |

[a] Exports minus imports.

*Sources:* South African Central Statistics Service, *South African Statistics Yearbook* (Central Statistics Service: Pretoria, various years); and South African Reserve Bank, *Quarterly Bulletin* (South African Reserve Bank: Pretoria, various years); and private communications with Armscor officials.

8 per cent both defence and civilian technology.[75] There is a perception among South African private-sector defence firms that international joint ventures and alliances provide the best possible route for survival and the growing number of collaborative programmes since 1994 strongly confirms this. Such patterns in the internationalization of defence production have, however, created new concerns about the global proliferation of conventional weapon systems[76] and could have normative implications for South Africa's arms trade decision making.

Despite the costs and problems associated with diversification during the transition period, most private-sector defence firms continued to pursue diversification strategies after April 1994. According to data supplied by SADIA, civilian sales for private-sector defence firms (including Altech, Grintek and Reunert) grew from 25 per cent of total sales in 1992 to 56 per cent in 1995. This trend was reflected in individual groups, such as Grintek, where the share of civilian business in total sales grew from around 16 per cent in 1992 to nearly 25 per cent in 1995.[77]

The significant increases in civilian business in 1994/95, for both the three major private-sector groups and the rest of the SADIA firms, were partly a function of firms' diversification efforts but were also related to the economic upturn which started in 1994 and gathered momentum in 1995, and which provided easier access to civilian markets, although South Africa's poor labour

[75] South African Defence Industry Association, *Some Statistics of the SADIA Members of the Defence Industry* (SADIA: Johannesburg, Jan. 1996), p. 8.
[76] Keller, W. W., *Arm in Arm: The Political Economy of the Global Arms Trade* (Basic Books: New York, 1995).
[77] Private communication with A. Holloway, Grinaker Avitronics, 13 May 1996.

**Table 7.13.** Arms industry employment, 1989–95

|  | 1989 | 1990 | 1991 | 1992 | 1993 | 1994 | 1995 |
|---|---|---|---|---|---|---|---|
| Armscor/Denel (th.) | 26.4 | 23.6 | 21.4 | 16.4 | 15.0 | 14 9 | 15.3 |
| Total arms industry (th.)[a] | 131.8 | 118.2 | 106.9 | 83.4 | 75.0 | 74.6 | 76.3 |
| % change on previous year |  | – 10.3 | – 9.5 | – 22.0 | – 10.1 | – 0.5 | 2.2 |
| Manufacturing employ't (th.) | 1 583.3 | 1 581.7 | 1 546.9 | 1 504.2 | 1 477.3 | 1 480.5 | 1 493.1 |
| Arms industry as share of manufacturing employ't (%) | 8.3 | 7.5 | 6.9 | 5.5 | 5.1 | 5.0 | 5.1 |
| Total employment (th.) | 5 650.2 | 5 633.3 | 5 537.5 | 5 424.8 | 5 312.2 | 5 278.4 | 5 317.8 |
| Arms industry as share of total employ't (%) | 2.3 | 2.1 | 1.9 | 1.5 | 1.4 | 1.4 | 1.4 |

[a] Includes all staff of private- and public-sector defence firms involved in both defence and non-defence work.

*Sources*: *Armscor Annual Report*, various years; *Denel Annual Report*, various years; and South African Reserve Bank, *Quarterly Bulletin* (South African Reserve Bank: Pretoria, various issues).

productivity and overvalued currency proved a significant obstacle to entry into international civilian markets.[78]

### Assessment of defence industrial adjustment since 1994

South Africa's arms industry continued to down-size and restructure. The most popular methods included cutting jobs, consolidating and rationalizing defence activities, increasing exports and international joint ventures and alliances with foreign defence firms. The overall size of the domestic arms industry, as measured by the total value of arms sales (exports and domestic sales) by domestic defence firms, continued to decline after 1994. In 1995 the total value of sales was 4 billion rand in current prices, over 25 per cent lower in real terms than in 1993. The fall in the value of total sales was largely a function of the dramatic declines in domestic defence expenditure, but the fall in the value of domestic sales was offset to some extent by real increases in the value of exports. The value of arms exports increased by 60 per cent in real terms between 1992 and 1995 while the share of exports in total defence sales increased from 12 per cent in 1992 to 25 per cent in 1995 (see table 7.10).

The real increases in the value of defence exports were not reflected in the share of defence goods in total exports and manufactured exports. Defence exports as a share of total exports remained constant at *c.* 1 per cent between 1993 and 1995, while as a share of manufactured exports they declined slightly to 2 per cent in 1995, down from over 3 per cent in 1993 (see table 7.12).

[78] GDP growth was 2.7% in 1994 and 3.3% in 1995, after being negative for the early 1990s. *South Africa Country Report, 2nd Quarter 1996* (Economist Intelligence Unit: London, 1996), p. 3. See also 'Economic growth: opportunity for prosperity', *Financial Mail*, 6 Oct. 1996.

**Table 7.14.** The structure of the defence market, 1992–95

|  | 1992 | 1993 | 1994 | 1995 |
|---|---|---|---|---|
| Total procurement expenditure[a] (m. rand, constant 1990 prices) | 3 243 | 3 162 | 2 427 | 2 167 |
| % change on previous year | − 22.8 | − 2.5 | − 23.2 | − 10.7 |
| Arms imports (m. rand, constant 1990 prices) | 618 | 644 | 421 | 446 |
| as % of total procurement | 19.0 | 20.3 | 17.3 | 20.5 |
| Domestic procurement expenditure (m. rand, constant 1990 prices) | 2 615 | 2 518 | 2 006 | 1 721 |
| Domestic as share of total procurement (%) | 81.0 | 79.7 | 82.7 | 79.5 |

[a] Includes procurement for the SADF, the police and other government departments.

*Sources: Armscor Annual Report*, various years; and South African Defence Industry Association, *Some Statistics of the SADIA Members of the Defence Industry* (SADIA: Johannesburg, Jan. 1996).

The declining size of the arms industry was also reflected in the trends in employment (see table 7.13). Employment in the public-sector arms industry bottomed out during 1994 and in 1995 Armscor and Denel together employed 15 250 people, slightly more than in 1994.[79] The share of the arms industry in total manufacturing employment was 5.1 per cent in 1995, down from over 8 per cent in the late 1980s.

The structure of the domestic defence market as measured by shares in procurement expenditure also saw significant changes during the same period (see table 7.14).

The share of imports did not increase significantly after 1993, despite the lifting of the UN arms embargo. Despite real declines in the value of Denel's domestic defence business, its share of the domestic market increased quite significantly after 1993, from 40 per cent to over 50 per cent in 1995. This was a function of its attempts to integrate vertically and subcontract less of its defence work. The private sector as a whole saw increases in its share of domestic procurement expenditure in 1992–93 before a decline in 1994 and 1995 (see table 7.15).

In addition to the changing size and structure of the domestic defence market, the contribution of the arms industry to the South African economy continued to decline after April 1994. The industry's total sales (including exports) of 4 billion rand (in current prices) in 1995 amounted for 1 per cent of GDP (1.5 per cent in 1989) and 4 per cent of total manufacturing output (6.6 per cent in 1989: see table 7.10).[80]

[79] Total employment among SADIA members in 1995 was estimated at 22 800, of which 16 600 were directly involved in defence work. *Some Statistics of the SADIA Members of the Defence Industry* (note 75).

[80] Estimates based on information from South African Reserve Bank, *Quarterly Bulletin* (various issues); and *South Africa Country Report, 2nd Quarter 1996* (note 78).

**Table 7.15.** The structure of domestic arms procurement, 1992–95

|  | 1992 | 1993 | 1994 | 1995 |
|---|---|---|---|---|
| Domestic procurement expenditure[a] (m. rand, constant 1990 prices) | 2 625 | 2 518 | 2 006 | 1 721 |
| % change on previous year | −21.7 | −4.0 | −20.3 | −14.2 |
| Denel domestic sales (m. rand, constant 1990 prices) | 1 391 | 1 069 | 932 | 907 |
| as % of domestic procurement | 52.9 | 42.4 | 46.4 | 52.7 |
| Private-sector sales (m. rand, constant 1990 prices) | 1 234 | 1 449 | 1 074 | 814 |
| as % of total domestic procurement | 47.1 | 57.6 | 53.6 | 47.3 |

[a] Includes procurement for the SADF, the police and other government departments.

*Sources: Armscor Annual Report*, various years; *Denel Annual Report*, various years; and South African Defence Industry Association, *Some Statistics of the SADIA Members of the Defence Industry* (SADIA: Johannesburg, Jan. 1996).

## VII. Conclusions

This chapter examines the adjustment experiences of the domestic arms industry since April 1994. In the context of declining defence expenditure, and in the absence of any clear government policy, the industry has continued to down-size and restructure. The lack of any clear policy direction from government towards the arms industry, largely as a result of the extremely time-consuming White Paper and Defence Review policy processes, has meant that restructuring and down-sizing have taken place in an ad hoc way and have been driven by market forces, with the result that valuable skills and technologies within the industry remain underutilized or are in the process of being lost.

The government's decision in early 1997 to produce a White Paper on the Defence Industry, while long overdue, has been welcomed by the arms industry and by most commentators and defence policy analysts. While the formulation of policy on the arms industry will not necessarily 'save' or prevent the further down-sizing of the industry, it should provide a coherent and consistent policy framework within which the industry can plan for the future.

The prospect of a smaller South African arms industry in the future is quite likely, given the possibility of further defence cuts in the context of macro-economic constraints and the fact that many of the SANDF's new equipment requirements (such as new jet trainers and corvettes) will be met by off-the-shelf purchases from overseas suppliers. Furthermore, the highly competitive international arms market and the pressures, from both within and without, for a further tightening of South Africa's arms export controls may constrain the industry's export drive, thereby limiting its ability to offset the impact of the declining domestic market.

# 8. Strategies for conversion at the national level

## I. Introduction

The legacy of South Africa's military burden and the socio-economic costs of the down-sizing of its defence industrial base present some difficult quandaries for the new South African Government. A crucial concern is the future size and shape of the defence industrial base. The adjustments taking place within South Africa's military sector also raise some fundamental policy issues linked to the socio-economic costs of disarmament, for instance, how to optimize the 'peace dividend', how to mitigate the negative impact of defence cuts, whether to adopt a national conversion strategy, how conversion fits into the broader transition process, and what the appropriate management structures for overseeing the change should be.

Drawing on widespread international experience, this chapter attempts to look at the range of policy options open to the South African Government that could help it to optimize the use of savings from defence cuts and to mitigate the industrial, technological and employment effects of defence expenditure cuts—in other words, how best to translate disarmament and demilitarization into an economic opportunity rather than a cost for South African society.

## II. The costs of disarmament

A starting-point for formulating policies is the identification of the groups or communities likely to bear the brunt of the effects of disarmament. Figure 8.1 below outlines the major sectors of society that are affected by the costs of disarmament. Once the relevant sectors have been identified questions arise about the extent of the effects, for instance, how much labour has been displaced from the armed forces and from arms industries and how much capital, land and technology and other inputs have been affected. Questions also arise about the regional distribution of the defence cuts.

In South Africa the deep cuts in defence expenditure led to the contraction of the military sector. Rationalization measures involved the closure of military bases (5 out of 12 air bases were closed between 1990 and 1993), the disbanding of various military units (such as the marines) and the loss of c. 37 000 armed forces personnel.[1] At the same time reductions in the procurement budget have had a dramatic impact on demand for indigenously produced equipment. Since 1990 at least 11 major weapon programmes have been cancelled, while another 49 have been reduced or postponed. As a result employment in the arms

---

[1] *South African Race Relations Survey 1994/95* (South African Institute for Race Relations: Johannesburg, 1995), p. 154; and appendix 1 in this volume.

industry has fallen from over 130 000 in 1989 to 76 250 in 1995. The rationalization and cost-cutting measures have had adverse affects on defence-dependent communities such as the naval port of Simonstown in the Western Cape and Gauteng Province (formerly the PWV region, renamed after the elections) where the majority of defence contractors are located. Added to the human suffering associated with labour displacement are the costs associated with the loss of expertise and technological capabilities as a result of the downsizing of high-technology facilities and the laying off of skilled defence workers. Evidence seems to suggest that the current restructuring of the arms industry has resulted in a number of highly skilled defence workers being enticed abroad by overseas defence firms, representing a net loss to the national technology pool.[2]

It is apparent that, no matter how desirable the military expenditure cuts may be, they are not without their problems. As previous chapters have shown, adjustment difficulties have been reflected in the collapse of defence market niches and in the failure of many small businesses indirectly dependent on military expenditure. The widespread expectation that defence expenditure cuts would provide an opportunity to revitalize the South African economy, or at the very least be used to aid those communities adversely affected by the downsizing of the arms industry, has been sorely disappointed. Instead the ripple effects of defence expenditure cuts have left a legacy of industrial and technological decline and economic hardship in communities dependent on defence.

As a consequence of these negative impacts the economic costs of disarmament have been perceived as a 'peace penalty' by those sections of South African society most adversely affected. Such perceptions are readily picked up by the South African defence lobby, who are quick to exploit the short-term negative impacts of defence cuts in their propaganda and to emphasize the role of the arms industry as 'a national asset economically, technologically and industrially'.[3] This statement begs the question, a strategic asset for whom? Certainly not for the majority of South African citizens, whose main experience of the products of the arms industry was on the front line in the townships. Moreover, given that there are few defence jobs for the black population, the defence of 'white' jobs does not hold much popular currency with the majority of the newly enfranchised population. For them maintaining a strategically important arms industry means denying resources for basic needs.[4] For many in South Africa today the costs associated with disarmament are the unfortunate legacy of apartheid's arms build-up. The new voices within the South African body politic are opposed to sustaining the defence industrial base and their prime concern has been the reallocation of resources for civilian purposes.

[2] Batchelor, P., 'Conversion of the South African arms industry: prospects and problems', Paper presented at the Military Research Group Seminar on Arms Trade and Arms Conversion in a Democratic South Africa, Pretoria, June 1993.

[3] South African Defence Industry Association, Submission to the Cameron Commission by the South African Defence Industry Association (Cameron Commission: Cape Town, June 1995).

[4] Bishop Storey, P., 'Moral and ethical issues raised by the defence industry', Paper presented at the Defence Industry Conference, Midrand, 11–12 Oct. 1995.

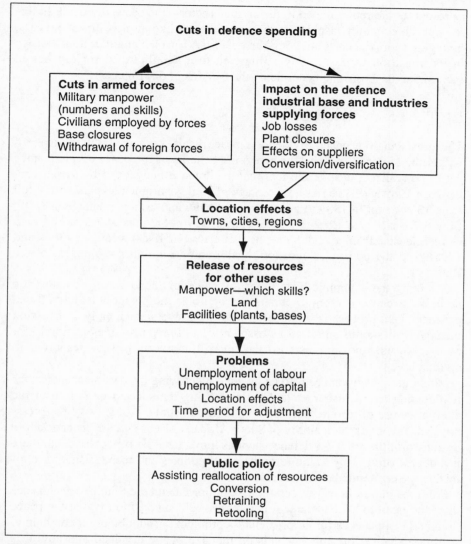

**Figure 8.1.** The costs of disarmament

*Source:* United Nations Institute for Disarmament Research, *The Economic Aspects of Disarmament* (UNIDIR: Geneva, 1993), p. 54.

Drawing on other countries' experiences it would appear that the duration of the negative impacts of defence expenditure cuts is usually short-term.[5] In the long term, disarmament can result in significant and beneficial gains through the expansion of production of civilian goods and services and/or social welfare

[5] United Nations Institute for Disarmament Research, *The Economic Aspects of Disarmament: Disarmament as an Investment Process* (UNIDIR: Geneva, 1993), p. xiii.

as resources are redistributed to the civilian sector—the peace dividend. In this respect the economic aspects of disarmament are like an investment process, involving short-run costs and long-run benefits.[6] The challenge for South Africa in its transition from a highly militarized to a demilitarized society lies in minimizing the costs and maximizing the benefits of disarmament.

## III. The peace dividend

The most tangible peace dividend is peace itself, both in South Africa and within the region. However, the economic benefits of the peace dividend have yet to be realized, despite the fact that the defence budget fell by 45 per cent between 1989 and 1995. As a percentage of total government expenditure it fell from 13 per cent in 1989 to 5.8 per cent in 1996 and, as a percentage of GDP, from 4.1 per cent in 1989 to 2.2 per cent in 1996.[7] Savings of 1.6 billion rand were released from the defence budget between fiscal years 1993/94 and 1995/96 to the housing, education, health and sanitation programmes of the RDP.

Realizing the economic benefits of the peace dividend is not an automatic process: it requires actors to pursue and optimize the potential benefits it can produce. This process is often referred to as conversion. Ideally, conversion management should entail an assessment of military resources available for alternative use and the costs associated with converting these resources to civilian use.

Much more attention needs to be given to identifying the best mechanisms for reallocating saved resources and for evaluating the success or failure of the alternative use of resources in the South African context. In view of the pressing social needs of the majority of South African citizens and the general failure to transform the 1989–95 defence savings into a tangible peace dividend, there is a degree of urgency to the formulation of policies and mechanisms for optimizing the economic benefits of disarmament.

From the outset, however, certain assumptions about the South African peace dividend need to be dispelled. A commonly held supposition is that the peace dividend is a panacea for all of society's problems. Undoubtedly it can help to facilitate change, but this depends on the size of the dividend and how it is used. Choices include whether it is used for private consumption, investment or social welfare or a mix of all three. If private consumption is the preferred option then the peace dividend may enable the implementation of tax cuts, putting more money into the consumer's pocket. If investment is preferred, it may be allocated to public investment in education and training (human capital formation) or infrastructure. On the other hand if the maximization of social welfare provision is a priority, resources are likely to be reallocated to social

---

[6] UNIDIR understands the peace dividend to be 'an investment process in which costs are incurred in the expectation of future benefits'. UNIDIR (note 5), p. xiii.

[7] See chapters 2 and 7 in this volume.

expenditure. Of all of these options investment is more likely to improve economic performance, particularly growth, than are allocations to social welfare payments or tax cuts.

Another popular misconception within the South African debate is the idea that the funds released from defence expenditure can be immediately transferred to other forms of government expenditure, as though the peace dividend were a fungible quantity of money which could be physically transferred from one place to another. This somewhat naive supposition ignores the fact that there is no automatic mechanism in society which reallocates resources and that conversion and/or the adjustment of military expenditure entail fundamental reallocations of resources in the economy, notably land, labour and capital, which by their very nature cannot readily be transformed without great dislocation and upheaval.[8]

## IV. Conversion

Various policy documents point to the direction the new government has chosen. The Reconstruction and Development Programme (RDP) stated that 'the democratic government must redirect military strategic production to civilian production'.[9] It goes on to argue that the defence force must be transformed from being an agent of oppression into an effective servant of the community with the capacity to contribute to reconstruction and development and that 'South African society must be thoroughly demilitarised'.[10] The RDP as it is presently formulated does not provide the policy framework for tackling the particular problems associated with defence industrial adjustment. There are, however, other signs that the government is moving towards the formulation of a national conversion policy. The 1996 White Paper on National Defence[11] is categorical in its support for a national conversion policy: 'In circumstances of diminishing domestic defence expenditure and falling global arms sales, the industry will be encouraged to convert production capability to civilian manufacture without losing the key technological capability needed for military production'.[12] This policy statement echoes the many calls from civil society to use the peace dividend in support of the conversion and/or diversification of the arms industry. Most notable were the large number of submissions to the Cameron Commission which called for the adoption of a national conversion strategy.[13]

---

[8] UNIDIR (note 5), pp. 67–69.

[9] African National Congress, *The Reconstruction and Development Programme: a Policy Framework* (Umanyano Publications: Johannesburg, 1994), p. 98.

[10] *Reconstruction and Development Programme* (note 9), p. 120.

[11] South African Department of Defence, *Defence in a Democracy: White Paper on National Defence* (Government Printer: Pretoria, 1996).

[12] White Paper on National Defence (note 11), p. 41.

[13] See, e.g., Pollecut, L., 'The Black Sash Submission to the Cameron Commission', Cameron Commission; Cock, J., 'Submission to the Cameron Commission'; and Crawford-Brown, T., 'Anglican Church Submission', Cameron Commission, Cape Town Public Hearings (June 1995).

The issue of defence industrial conversion was explicitly raised in the 1994 Transitional Executive Council Sub-Council on Defence discussion document on a national policy for the arms industry.[14] Alongside the argument for retaining strategically important segments of the arms industry the authors asserted a clear need to implement a policy of diversification 'that encourages and assists defence companies to utilize their potentially multi-purpose (military/civilian) capabilities to meet civilian market needs'.[15] Beyond such rhetoric, clear policy guidelines on defence industrial conversion have yet to emerge. Defence policy reformulation is inching its way through a number of commendably open and transparent review processes. In the meantime the issue of conversion and diversification is being hotly debated in a number of public arenas, as the various interest groups hone their arguments for and against.

To many of the active groups within South Africa's civil society that are concerned with military issues, conversion is seen as an essential element of the demilitarization process, specifically designed to dismantle the militarized sections of the industrial base that were built up during the Total Strategy era. As Rogerson comments, 'the question of defence industrial adjustment looms large on the policy horizon of the post apartheid reconstruction'.[16]

For a semi-industrialized developing country like South Africa there are considerable opportunity costs associated with maintaining a relatively large, modern, technologically advanced defence force, not least because the rate of inflation for military equipment tends to be higher than the rate of inflation for civil goods, creating substantial upward pressures on the defence budget over time. In the past the South African government was able to maintain such inventories because the inequalities of the apartheid system ensured that huge subsidies could be provided to the defence sector at the expense of the living standards of the vast majority of South Africa's population. It is not in the interests of the ANC to be associated with the perpetuation of this model of militarism.[17]

## The international experience

Much of the inspiration for the domestic debate has been drawn from contemporary international experience. This is a healthy sign, for, as UNIDIR argues, 'the access to and the use of foreign experience and knowledge in the development of civilian production is a key factor for successful conversion for all countries, especially in countries where the civilian sector of the economy has been neglected and suppressed by military priorities and claims'.[18] The issue of defence industrial conversion has occupied the minds of policy makers and

[14] Transitional Executive Council, Sub-Council on Defence, National Policy for the Defence Industry (Subcouncil on Defence: Pretoria, Apr. 1994).

[15] National Policy for the Defence Industry (note 14), p. 4.

[16] Rogerson, C., 'The question of defence conversion in South Africa: issues from the international experience', *Africa Insight*, vol. 25, no. 3 (1995), p. 154.

[17] Willett, S. and Batchelor, P., 'The South African defence budget', Military Research Group, Johannesburg, 1994, p. 4. Unpublished mimeograph.

[18] UNIDIR (note 5), p. xiv.

government officials in many countries experiencing the effects of disarmament following the end of the cold war.

Since the end of the cold war the interest in conversion has grown as many countries have faced the costs associated with disarmament. At the same time the concept of conversion has become hotly contested, becoming an all-encompassing concept open to many interpretations.[19] A simple definition of conversion is the shift in resource use from military to civilian—swords into ploughshares or tanks into tractors. For example, Renner states: 'conversion encompasses an adaptation of research, production and management practices in arms producing factories to civilian needs and criteria'.[20] This somewhat simplistic and static view of conversion fails to capture the complexities of the process being proposed. In a sense the 'swords-into-ploughshares' approach misses the point made by the Bonn International Centre for Conversion that the goal of conversion is to maximize not the amount of resources converted but rather the overall benefits for the citizens or society.[21] This point is particularly significant in the case of a developing society like South Africa where the benefits of militarism accrued to a small élite at the expense of the majority of the population.

This approach also gives little recognition to the fact that the conversion of defence plants, or indeed military bases, is not always practicable or desirable. It may, for instance, make more sense to scrap the machinery from arms-producing plants if their running costs are higher than the costs of those devoted exclusively to civilian production. In the case of military bases, toxicity of the land may make its conversion to civilian use hazardous and expensive.

For some conversion analysts the issue of conversion is more specifically focused on the question of labour and skills.[22] Paukert and Richards, for instance, argue that the challenge posed by conversion is 'to make the best use of all the resources released by defence cuts, particularly the previously employed labour force, at the least social cost'.[23]

The 1993 UNIDIR report *The Economic Aspects of Disarmament: Disarmament as an Investment Process* defines conversion in two ways. First:

There is the narrow interpretation of converting defence plants into establishments manufacturing civil goods, and vice versa, converting civil plants into establishments manufacturing military goods. Effectively this interpretation requires product substitution in which the same plant and workforce produces civilian products instead of military products. Such product substitution is not possible for the armed forces remaining

[19] Adelman, K. and Augustine, N., 'Defence conversion', *Foreign Affairs*, vol. 71, no. 2 (1992), pp. 26–47; and Cooper, J., 'Conversion is dead, long live conversion', *Journal of Peace Research*, vol. 32, no. 2 (1995), pp. 129–32.

[20] Renner, M., *Swords into Ploughshares: Converting to the Peace Economy*, Worldwatch Institute Paper no. 96 (Worldwatch Institute: Washington, DC, 1992).

[21] *Conversion Survey 1996: Global Disarmament, Demilitarisation and Demobilisation* (Bonn International Centre for Conversion: Bonn, 1996), p. 20.

[22] Paukert, L. and Richards, P. (eds), *Defence Expenditure, Industrial Conversion and Local Employment* (International Labour Office: Geneva, 1991).

[23] Paukert, L. and Richards, P., 'Employment impacts of industrial conversion', eds Paukert and Richards (note 22), p. 205.

within the military sector. However, with a change of ownership, there are product substitution possibilities for some of the armed forces defence facilities, which can be converted into civilian uses. Military air bases, for example, can be used as civil airports.

The second interpretation of conversion uses a broader notion by focusing on 'the process of reallocating resources released from the armed forces and from declining defence industries to the expanding sectors and regions of the economy. This factor reallocation process is occurring continuously in any dynamic economy and its success depends on the operation of the markets for labour [and] capital and on the general state of the economy'.[24]

Other analysts are unhappy with the term 'conversion', preferring to use other words such as 'diversification' or 'adjustment strategies', while others argue that conversion is impossible to achieve in practice. Some conclude that conversion has failed while others argue that it is unwarranted.

The contention over the definition of the concept of conversion strengthens Harbor's point that there can be 'no universal plan or blueprint for arms conversion'. He warns of the dangers of trying to adopt 'off-the-shelf' strategies for conversion.[25] Policy responses differ according to a number of factors operating at national, industrial/sectoral and corporate levels of investigation and analysis. Accordingly successful strategies for conversion need to 'be based on local economic, technical and political realities and circumstances, and will encompass a range of policies to deal with separate goals and problems'.[26] In much the same vein Renner contends that:

each country must chart its special path toward a peace economy, judging by the degree of government involvement desirable, deciding how much leeway to give communities and regions, determining whether conversion should be targeted toward expanding the supply of consumer goods or providing 'socially useful' products and services like health care and education programs, public infrastructure, and environmental protection.[27]

Despite this appeal for nationally specific conversion approaches, it has been a useful exercise for South African policy makers to identify the major threads within the international conversion debate, if for no other reason than to avoid the pitfalls experienced by other countries.

### Approaches to conversion

From the existing international literature it is possible to distinguish three distinct approaches to conversion: (a) macroeconomic, (b) micro-economic,

[24] UNIDIR (note 5), pp. 52–53.

[25] Harbor, B., 'Lessons from the West European Arms conversion experience', Paper presented at the Military Research Group Seminar on Arms Trade and Arms Conversion in a Democratic South Africa, Pretoria, June 1993.

[26] Harbor (note 25), p. 3.

[27] Renner (note 20), p. 8.

including company, plant-based and community- or region-based approaches, and (c) political conversion.[28] In varying degrees all these approaches have their advocates within the South African debate.

A distinction must be drawn here between enterprise conversion and diversification, which are often used interchangeably but are actually quite different. Conversion means the transformation of military production facilities into the production of civilian goods and implies that all or part of a defence firm's defence activities is transformed into civil production. Diversification, on the other hand, involves a company shifting into other areas of production, either through internal investment or through takeovers and mergers, without converting defence industrial facilities. Thus diversification does not necessarily imply product substitution: rather it reflects market substitution, that is, moving into new markets. This has the advantage of making a defence contractor less vulnerable to fluctuations in demand for defence goods, and may even enable a firm to shift into other defence areas, with all the subsequent capability to facilitate a rearmament process, if there is an upturn in the defence market.

Macroeconomic studies of conversion concern themselves with the demand side of the problem, analysing the impact on the economy of defence cuts and reorienting public expenditure to other categories of government expenditure, often without considering the problems of defence adjustment, or simply assuming that the market will deal with them. In contrast the micro-economic approaches focus on the supply-side adjustments required at the level of the company, plant and community. They provide an alternative to the market-oriented approach, which assumes that the market will provide the right signals for firms to reduce defence production and increase production of civilian goods and services. Political conversion, on the other hand, attempts to deal with both the supply and the demand side of the dynamic in a coherent and consistent manner, and has the explicit goal of demilitarization.

*Macroeconomic conversion*

The macroeconomic approach focuses on the evidence of the negative relation between military expenditure and economic development. Melman and Smith were pioneers of this perspective.[29] While there is still some debate, as discussed in Dunne,[30] it is generally accepted that military expenditure represents an economic burden and that reducing it and transferring resources to investment in the civil sector can improve economic performance. Moreover, these resource transfers would create higher levels of employment, as investment in the civil sector has been shown to create more employment than investment in

[28] Dunne, P. and Willett, S., *Disarming the UK: The Economics of Conversion*, Discussion Paper Series (Centre for Industrial Policy and Performance, University of Leeds: Leeds, 1992).

[29] Melman, S., *The Permanent War Economy: American Capitalism in Decline* (Simon and Schuster: New York, 1974); and Smith, R., 'Military expenditure and capitalism', *Cambridge Journal of Economics*, vol. 1, no. 1 (Mar. 1976). See also chapter 1 in this volume.

[30] Dunne, P., 'The political economy of military expenditure: an introduction', *Cambridge Journal of Economics*, vol. 14, no. 4 (Dec. 1990), pp. 395–404.

the military sector.[31] At the core of this approach is the focus on macroeconomic adjustment policies. The government can use defence savings to reduce the budget deficit, to increase other government expenditures (current and capital) or to cut taxes.

In a macroeconomic study conducted by the IMF, which addressed the global economic effects of reduced military expenditure, the impact of a 20 per cent cut in military expenditure over an 11-year period was predicted to be a 3.9 per cent annual reduction in government consumption, a 1 per cent annual increase in private consumption, a 1.8 per cent increase in private investment and a 0.3 per cent annual increase in GDP in developed economies. The results for developing countries were even more positive.[32] Notwithstanding the IMF results, the way in which the peace dividend affects an economy is equivocal, depending on political choice and national preference.

Barker *et al.* and Gleditsch *et al.* provide examples of macroeconomic conversion.[33] The evidence from these studies suggests that at a macroeconomic level there should be no problems in adjusting to lower levels of military expenditure and that with sensible policies of adjustment there can be improved economic performance. In other words there should be a tangible peace dividend. These studies also discuss the context of the changes and some of the likely problems of adjustment. In particular they present estimates of the likely industrial impact and suggest where there may be adjustment problems.

To a great degree the RDP exemplifies the macroeconomic approach in the South African context. It was largely conceived as a mechanism for redistribution aimed at transforming national priorities and meeting the basic needs of the majority of the population. In this spirit defence savings were consciously redirected towards the RDP from 1994 onwards.[34] In addition to the human capital investment that the RDP embodies, it was anticipated that these programmes would generate a considerable amount of employment once they were up and running.

The aspirations of the RDP are tailored to the goal of overcoming the inequities of the apartheid system, and are thus an essential precondition for the genuine transformation of South African society, but they do not address the specific task of demilitarizing the economy or the defence industrial adjustment problems that this entails. Evidence from many countries shows that the effect of the macroeconomic redistribution of defence savings is often structural unemployment of displaced defence workers.[35] This is largely because there has been no conscious policy of reallocation. The displacement of defence workers

[31] Melman, S., *From Military to Civilian Economy: Issues and Options*, Occasional Paper Series no. 8 (Centre for the Study of Armament and Disarmament, California State University: Los Angeles, Calif., 1981).

[32] International Monetary Fund, *World Economic Outlook 1993* (IMF: Washington, DC, 1993), pp. 104–112.

[33] Barker, T., Dunne, P. and Smith, R., 'Measuring the peace dividend in the United Kingdom', *Journal of Peace Research*, vol. 28, no. 4 (Nov. 1991), pp. 345–58; and Gleditsch, N., Cappelen, A. and Bjerkholt, O., *The Wages of Peace: Disarmament in a Small Industrialized Economy* (Sage: London, 1994).

[34] Willett and Batchelor (note 17).

[35] For detailed studies, see Paukert and Richards (note 22).

represents a loss of knowledge and skills from a nation's technological pool, and thus raises particular concern in terms of its implications for the economy's future technological performance. Given the technological importance of the South African defence sector it has been necessary for South Africans to examine other approaches to conversion which are explicitly geared to dealing with the problems of defence industrial and technological adjustment.

## Micro-economic conversion

Micro-economic conversion includes enterprise- and/or plant-based conversion and is focused essentially on the reuse of existing military resources for civilian purposes. The main characteristic of this approach is that the decision to diversify is made at the enterprise level.

In the recorded experience of South African defence contractors, conversion has not been successful and diversification has been preferred. It can occur either in the form of mergers and acquisitions or by the strengthening of the civil operations of the enterprise. In South Africa several defence contractors have initiated policies of diversification. Within the public-sector company Denel, for instance, several attempts have been made to produce commercially viable civilian products (see table 8.1).

There have also been several attempts at private-sector diversification, as highlighted in chapter 5. Despite these encouraging examples, their significance is marginal in terms of the companies' overall activities, and they by no means compensate for the scale of the rationalization taking place in their defence divisions. In general the majority of defence contractors have displayed a reluctance or inability to diversify from the start of the downturn in defence expenditure. This can partly be explained by the severe recession in the early 1990s which created a disincentive to invest when demand was depressed in all major civil markets.

International evidence for the successful conversion of existing plants to civil production is extremely thin, notwithstanding some isolated although well publicized cases in China, Russia and Ukraine.[36] Dussauge has developed a simple matrix that can be used to evaluate the possibilities for converting industrial capacity from military to civil work (see figure 8.2). Firms in section A are the ones in which conversion is most likely: this is because they are usually involved in the production of generic technologies which have dual-use applications. In contrast, firms in section D confront the greatest obstacles to conversion because defence sales dominate their business and they depend on defence-led technology which is highly specific to the defence market.

---

[36] Samorodov, A., 'Conversion in the Soviet defence industry and its consequences for labour', *International Labour Review*, vol. 129 (1990), pp. 555–72.

**Table 8.1.** Denel: diversification and conversion efforts

| Denel subsidiary | Military products | Civilian products |
|---|---|---|
| Swartklip | Pyrotechnics<br>Ammunition<br>Explosives<br>Artillery shells | Explosives equipment for mining and<br>  construction industries<br>Emergency flares |
| Naschem | Medium and heavy-calibre<br>  ammunition | Explosives for mining industry |
| Somchem | Artillery charges<br>Rocket systems<br>Missiles (air-to-air)<br>Propellants<br>Mortars | Nitro-cellulose and chemical products<br>Composite materials<br>Medical equipment<br>Reinforced polyester pipes |
| PMP | Small- and medium-calibre<br>  ammunition | Brass/copper products<br>Pressed/machined products for motor<br>  industry<br>Drill bits<br>Explosive bonding |
| Atlas Aviation | Military aircraft<br>Military helicopters | Commercial aircraft<br>Engine gearboxes |
| LIW | Automatic cannons<br>Small arms (pistols, rifles)<br>Artillery systems<br>Mortars | Skid-steer loader<br>Mining equipment<br>Machining |
| Kentron | Missiles<br>Air defence systems<br>Remotely-piloted vehicles<br>Avionics<br>Stand-off weapons | Turnkey security systems<br>Traffic engineering systems<br>Observation systems |
| Eloptro | Electro-optical equipment<br>Sights<br>Infra-red | Thermal imaging<br>Laser products<br>Night vision products |
| Musgrave | Rifles | Sporting rifles<br>Cricket bats |
| Houwteq | Military spy satellites | Observation satellites |

*Source: Denel Annual Report*, various years.

In contrast to company-based strategies which centre on management deci-sion making, plant-based conversion has emerged from workers' responses to job losses in the arms industry and is based on the classic notion of 'swords into ploughshares'.[37] It is concerned with the transformation of resources which have already been accumulated by the military, namely defence production facilities, skills and scientific and technical knowledge and in some instances

---

[37] For a South African trade unionist view of the defence industry and conversion, see Fanaroff, B., 'A trade unionist perspective on the future of the armaments industry in South Africa', *South African Defence Review*, no. 7 (1993 ), pp. 10–14.

| Asset specificity[a] | Dependence on defence sales | |
| --- | --- | --- |
| | Low | High |
| Low | A (relatively easy) | B |
| High | C | D (most difficult) |

**Figure 8.2.** Prospects for conversion

[a] Asset specificity is defined as defence-led and defence-specific technology.

*Source:* Dussauge, P., 'The conversion of military activities', eds F. Blackaby and C. Schmidt, *Peace, Defence and Economic Analysis* (Macmillan: London, 1987).

military products and components. The most renowned attempt at plant-based conversion was the Lucas Corporate Plan produced by British defence workers in 1975–77.[38] Central to the Lucas Plan was the notion of reshaping technological priorities through the development of 'socially useful products' which could be made utilizing the existing resources and skills of defence plants. Although never implemented, the Lucas Plan provided a model for many workers' plans and workplace conversion committees.[39]

At the centre of the plant-based approach to conversion is the alternative use committee. This committee should ideally include management, labour and community representatives who would join forces to draw up an alternative business plan.

While some interest in alternative use committees has been expressed among South African trade unionists, local authorities and peace groups, so far no such committees have been set up. This may be due to the general lack of success in other countries where alternative use committees have experienced a lack of cooperation from management and many other barriers to plant-based conversion. Not least importantly, the social groups which make up alternative use committees do not always seek the same objectives. The priority for defence workers and the trade unions is to retain jobs and as a last resort this will mean continuing to lobby for defence contracts, which grates against the objectives of peace activists. Peace activists, on the other hand, with their high moral commitment to achieving disarmament have often been insensitive to the plight of defence workers and their families.

In general, alternative use committees have operated under unrealistic assumptions about how business and markets function in the commercial world. The lack of technical expertise and knowledge of the innovation process may

[38] Wainwright, H. and Elliott, D., *The Lucas Plan* (Allison and Busby: London, 1982).

[39] Other plant-based conversion campaigns in the UK during the 1980s were centred on the Llangenech Naval Depot, British Aerospace Dynamics at Bristol, the Enfield Small Arms Factory in outer London and VESL in Barrow, Cumbria. Willett, S., 'Conversion policy in the UK', *Cambridge Journal of Economics*, vol. 14, no. 4 (1990), p. 476.

lead alternative use committees to make technical proposals with little realistic consideration of the costs of production, market opportunities or the required investment capital. Moreover, the specific features of defence firms such as their capital and labour endowments are rarely appropriate for volume commercial production. Defence prime contractors tend to produce custom-built products to a very high standard of specification, with an emphasis on the technological characteristics rather than costs. This is because the market is guaranteed and the customer (that is, the government), not the producer, bears the brunt of costs for product enhancement. Production runs are short and the production process is skill- , labour- and capital-intensive. In contrast, the production of all but the most luxurious of commercial goods is characterized by economies of scale and by mass production techniques which have a low labour : capital ratio and are driven by unit-cost considerations. Because governments guarantee both the markets and profits for defence firms, the production and labour processes within defence plants are often antiquated in comparison with those employed in commercial enterprises. In addition commercial companies have widespread experience in marketing and selling in highly competitive markets and dealing with a multiplicity of customers. Defence companies or divisions, in contrast, are used to dealing with one major customer and have built up the networks and experience to deal with what is essentially a highly bureaucratized market.

Kaldor has argued that plant-based conversion initiatives which seek product solutions to retaining defence skills and facilities are likely to replace the military–industrial complex with other complexes (such as a health complex) which have the same characteristics—the ability to extort funds from the government and produce grandiose products.[40] The attempt to produce civil products within the context of production structures which have evolved out of the military production and procurement system invariably reproduces the same technological characteristics of defence systems, namely products which are capital-intensive, complex, over-specified and costly. As such they reinforce the present industrial and technological patterns of defence production rather than transform them, representing a missed opportunity in terms of the potential for economic restructuring.

The emphasis in South African conversion programmes on product solutions to the underutilization of capacity and employment problems reflects a poor understanding of the process of innovation, technological change and market penetration. Conversion policies tend to ignore completely the innovation implications of the diffusion of existing process technologies. The great technological successes of Japan resulted not so much from the evolution of radical new products as from the application of existing techniques and technologies to a wide range of existing products, that is, digitalizing televisions, miniaturizing products through the use of microchips, improving the productivity of tradi-

[40] Kaldor, M., 'Problems of adjustment to lower levels of military spending in developed and developing countries', Paper prepared for the World Bank's Annual Conference on Development Economics, Washington, DC, 25–26 Apr. 1991.

tional industries through the use of flexible manufacturing systems, and applying information technologies which have given rise to 'just-in-time' production, all contributing to the highly competitive nature of Japanese products.[41]

Lower down the defence product hierarchy, those contractors producing generic technologies which have dual-use applications are unlikely to need to convert their production processes or products. For this group of contractors the crucial issue is to identify alternative markets for the intermediate goods they supply, which is dependent on healthy demand from civil industries. In this context a demand stimulus is far more important than supply-side adjustment strategies.

Last in the category of micro-economic conversion is the local community approach. This approach focuses on the impact of defence cuts on defence-dependent communities and emphasizes the role of local authorities and activist groups like trade unions and peace groups working together to develop strategies to reduce the vulnerability of local defence-dependent communities to fluctuations in defence expenditure. This approach has largely found its inspiration from the USA where local state initiatives supported by the Department of Defense's Office of Economic Adjustment were successfully implemented following the downturn in military expenditure at the end of the Viet Nam War.[42] Beyond a small number of successes in transforming defence facilities to civil production, the recent record of local and community-level initiatives has not been inspiring. The disappointment associated with such initiatives has led to a shift in emphasis from the plant-based conversion approach towards an approach which is linked to strategies of regional regeneration, inspired by the archetypal small-firm 'industrial districts' of Western Europe.[43] In this shift to broader local economic restructuring, some local conversion initiatives are beginning to intermesh with political conversion which is based on the principles of national and regional economic and social regeneration.

*Political conversion*

Political conversion is a more all-encompassing concept than those described above in that it simultaneously addresses the economic problems produced by demilitarization, demobilization, reduced defence expenditure and reductions in arms production.[44] In the context of demilitarization conversion implies a general shift in national priorities, presupposing the reallocation of resources from the military to the civil economy. A key factor in this strategy is the reassertion of civilian control over militarized political and economic institutions.

The political conversion approach emphasizes the way in which military imperatives dominate national technological and industrial objectives of high

---

[41] Best, M., *The New Competition: Institutions of Industrial Restructuring* (Polity Press: London, 1990).

[42] Rune, T., *Federal Response to Economic Crisis: The Case of Economic Adjustment* (Department of Defence: Washington, DC, 1977).

[43] Best (note 41).

[44] This approach has been developed by Dunne and Willett (note 28).

military spenders and defines a trajectory of economic development. Its proponents argue for the transformation of resources tied up in defence production as part of a broader social and economic agenda of industrial and technological reconstruction rather than as an end in itself. Central to this approach are the recognition of 'market failure' in advanced capitalist societies in meeting basic human and environmental needs and the urgent requirement of shifting industrial, technological and scientific resources away from military-defined objectives and instead targeting national needs like industrial development or renewal, environmental restoration, sustainable development and renewable energies. Such 'national needs' policies, although initiated by government in the first instance, should operate in partnership with industry, finance, local and regional authorities, trade unions and consumers.

In this context a conversion policy driven by a national needs agenda would require a broadly-based initiative aimed at using the peace dividend to tackle the full range of social and environmental problems confronting the country and a reorientation of industrial policy aimed at reinvigorating the flagging manufacturing sector.[45] Established industries such as mining and textiles might be targeted for support for modernization, but at the same time as new fledgling industries, particularly in high-technology growth areas such as information technology, were encouraged to spread their wings.

The mechanisms available to government to encourage such developments include direct subsidies, new regulations, particularly in the environmental field, and incentives for industries to conduct research, improve technology transfer and support job training and technical education. However, bringing a country's industrial and technology policies into line with its national needs requires more than drawing up a shopping-list of social and economic needs and then shifting funds from defence towards other public expenditure programmes. More fundamentally the country requires a new framework for guiding R&D policies, scientific and technological priorities and the relationship between the private and public sectors. Such a framework is required to provide a filter for selecting critical industries and technologies that would receive government support.

This approach has great saliency in South Africa because of the attempt to challenge the economic and industrial manifestations of militarism from the apartheid era which remain embodied within the defence industrial base and its support structures such as Armscor.

**Other options**

From the experience of defence companies both within South Africa and in other countries it is apparent that the norm for defence contractors facing an

45 Willett (note 39); Cronberg, T., 'The social reconstruction of military industries and technologies: seen from a Danish perspective', Paper presented at the International Institute of Peace Conference on Conversion, Vienna, Feb. 1992; and Yudkin, J. and Black, M., 'Targeting national needs: a new policy for science and technology', *World Policy Journal* (spring 1990).

unfavourable market environment and declining defence market is not to convert but to pursue offensive adjustment strategies such as increasing their monopolies in defence market niches, through mergers and acquisitions, the maximization of arms exports and joint ventures and alliances, while shedding those defence division that are no longer lucrative. This trend has been confirmed in the case of Denel subsidiaries such as LIW, Naschem and Pretoria Metal Pressings which have all increased their foreign sales since 1989, as have many private-sector defence firms. There is, however, a limit to this survival strategy dictated by present trends within the international arms market.

The value of global exports of major conventional weapon systems declined by 50.6 per cent between 1987 and 1994.[46] Scarce hard currency reserves in developing countries and the expansion of indigenous arms-production facilities have contributed to the downward trend in demand. As a consequence of arms control agreements and disarmament in Europe, many governments are unloading surplus arms onto the international market. Arms-producing companies are often in direct competition with these surplus weapons for their sales. At the same time the global arms industry is undergoing a profound transformation which in turn is affecting patterns of demand. The internationalization of arms production is replacing traditional patterns of production based on nationally owned and controlled arms industries. Transnational companies are emerging, fostered by the process of cross-border mergers and acquisitions, joint ventures and cross-share ownership. Domestic responses to defence industrial adjustment issues are important, particularly with respect to jobs, but it is clear from the process of internationalization that the conversion issue now has an international dimension.[47] Companies may avoid conversion through internationalization and in South Africa there are clear signs that this is what many defence firms are trying to do. The Atlas–Airod link-up to produce the Rooivalk helicopter is the obvious case in point.

The general failure of defence companies to adjust to the changing market environment arises from the specific characteristics of defence enterprises, most notably the culture and management style of the arms industry, no matter what its global location.[48] As Markusen has observed in the USA, the defence industrial base evolved a type of production process conspicuously different from that employed in commercially oriented plants, thus sustaining a whole section of industry dedicated to designing and producing ever more costly products in an environment which was highly inflexible, experimental and shrouded in secrecy.[49] This production structure has specific features which militate against

[46] Calculated from SIPRI data. Anthony, I. *et al.*, 'The trade in major conventional weapons', *SIPRI Yearbook 1997: Armaments, Disarmament and International Security* (Oxford University Press: Oxford, 1997), appendix 9A, p. 292.

[47] Laurance, E. and Wulf, H., *Conversion and the Integration of Economic and Security Dimensions*, BICC Report 1 (Bonn International Centre for Conversion: Bonn, 1995), p. 15.

[48] Alexander, A., 'National experiences in the field of conversion: a comparative analysis', Paper presented at the UN Conference on Conversion, Moscow, 13–17 Aug. 1990.

[49] Markusen, A., 'The military industrial divide', *Environment and Planning: Society and Space*, no. 9 (1991), pp. 391–416.

successful conversion. These features include inexperience in commercial marketing, emphasis on product rather than process innovation, a hierarchy of skills and a technological orientation structured by the weapon systems procurement process, a management organization isolated from commercial practices, a production structure defined by large-scale systems integration, protected markets through government procurement, and risk aversion as a result of government subsidies for R&D and capital investment.[50] The fact that the arms industry is used to a single buyer ordering a highly specialized product has made it reluctant and ill-placed to compete in civil markets. These factors coalesce to erect high barriers to exit from the arms industry and explain the limited degree of success in attempts to convert military enterprises.[51]

## Conversion and industrial strategy

Bringing the country's industrial and technology policies into line with its national needs has found expression in the objectives of the Industrial Strategy Project (ISP), which argues that the key to South Africa's long-term economic well-being lies in the reinvigoration of the manufacturing base.[52] It emerged from an initiative of COSATU in the late 1980s when it commissioned a group of economists to prepare a report analysing the impact of sanctions on the South African economy. The research revealed that the crisis of the South African economy was rooted in the policies of the apartheid era. A key consequence of the failures of apartheid's social and economic policies was an unproductive manufacturing sector. Appreciation of this problem inspired COSATU to encourage its research group to formulate a new industrial policy for South Africa. This resulted in a set of concrete proposals. Within the context of meeting the basic needs of the population the ISP highlights four interlinked objectives of industrial policy: (a) creating employment; (b) increasing investment; (c) raising productivity; and (d) improving trade performance.[53]

The link between conversion, industrial and technological restructuring and national needs was most keenly expressed at a workshop on the South African arms industry held in Johannesburg in February 1996 where calls were made to link defence industrial adjustment to a broadly-based industrial strategy designed to encourage the development of the South African economy. The strong trade union representation argued that the restructuring of the arms industry should be part of the more general reprioritization of South Africa's develop-

---

[50] Willett, S., 'Policies for adjustment to lower military spending', Paper presented at the Military Research Group Seminar on Arms Trade and Conversion in a Democratic South Africa, Pretoria, June 1993.

[51] Gansler, J., *Defence Conversion: Transforming the Arsenal of Democracy* (MIT Press: Cambridge, Mass., 1995), pp. 122–28.

[52] For details of this approach, see Joffe, A. *et al.*, *Improving Manufacturing Performance in South Africa: Report of the Industrial Strategy Project* (University of Cape Town Press: Cape Town, 1995).

[53] Joffe *et al.* (note 52), p. 16.

ment goals.[54] The proponents of this approach argue that a defence industrial adjustment strategy should target employment creation as its primary principle. This demand forms part of a general awareness among trade unionists and peace activists alike that economic insecurity caused by mass unemployment is the greatest threat to South Africa's long-term security and economic future in this century. They argue that the arms industry can contribute to this process because of: (*a*) its concentration of engineering skills, which are transferable to other engineering industries; (*b*) its fixed capital assets, machine tools, computers and so on; (*c*) its science and technology know-how; and (*d*) the potential for environmental conversion to address the pressing environmental problems such as acid raid due to over-use of solid fuels, water pollution, environmental clean-up of other industries and alternative transport systems.

To achieve these gains and minimize the skill losses from the defence sector, it was argued that the South African state will need to support such transfers. A number of mechanisms were identified in order to achieve these goals such as the reorientation of national R&D strategies and support for marketing new products or processes. However, the workshop members recognized that adjustment strategies need to be mindful of the government's budgetary constraints and must be developed with moderate initial investment risks. Because of the structural barriers to conversion within the defence sector, it was argued that mechanisms such as retraining, start-up incentives, tax breaks, civil R&D programmes and investment programmes should be designed to restructure capital and labour away from the defence industrial base and into more dynamic and innovative forms of production and labour processes, creating a radical break with the pervasive culture of the South African defence industrial base. The use of such mechanisms need not be prohibitively expensive and may offer more positive benefits to the whole of society in the long run.

Furthermore, it was argued that an industrial policy driven by national needs should encompass a broadly based initiative aimed at improving the full range of manufacturing industries, including those that produce goods needed for dealing with the pressing social and environmental problems confronting South African society. In this way a number of fledgling industries like solar cells or recyclable plastics could be targeted. So, however, should selected basic manufacturing industries such as the extractive and service industries, most of which receive little government R&D support but which are important to a balanced economy. At the same time as mature industries are being revived, several new 'strategic' high-technology civilian industries—microelectronics, computers, telecommunications, advanced materials, robotics and numerical controls—could also be targeted as they are crucial to the performance of other industrial sectors. These industries would no longer be dominated by military objectives and R&D policies directed towards these sectors would instead reflect commercial as well as social objectives.

---

[54] Report from the Industrial and Technology Strategy Workshop on the South African Arms Industry held by the Centre for Conflict Resolution and the Group for Environmental Monitoring, Johannesburg, Feb. 1996. Unpublished mimeograph.

Concerns were expressed, however, about the capacity of the South African state to manage such a strategy. This concern is evidenced by the problems with reprioritization within the RDP and the strong vested interests within the government, which allow for back-door budgeting in support of the arms industry. Fears were also expressed that defence industrial policy was being developed in isolation from other policy forums within South Africa. This was thought to be a direct result of the lobbying of the government by the arms industry in its attempt to pre-empt more open and transparent debates on defence industrial policy. In the light of these trends it has been recognized that there needs to be a much broader process of consultation over defence policy than at present—a process which would include stakeholders such as organized labour and concerned members of the public. Above all there is the recognition of the need for a public awareness campaign to empower civil society where defence industrial issues are concerned so that it can effectively contribute to the national debate on defence policy and defence industrial issues.

## V. Refocusing R&D

One requirement of defence industrial adjustment that is crucial if industrial reinvigoration is to take place is a major reform of the national science, engineering and technology (SET) base, particularly in view of the fact that South Africa's innovation strategies have been traditionally dominated by military concerns. The Green Paper on Science and Technology released in January 1996 observed that 'a substantial part of government-sector SET resources has been channelled to military and nuclear R&D with negligible benefits for the civilian population'.[55] Traditionally, a small and unaccountable bureaucratic élite established technology and R&D policies and allocated scarce public resources to SET objectives which were defined by the needs of the highly militarized pariah state. Civil institutions played a very limited and diffuse advisory role in the formulation of R&D policies. In view of this situation there is clearly a need for a national innovation strategy which reorients national technological priorities away from military-defined needs towards civil objectives.[56]

The Green Paper argued that the need for reform was most urgent in the light of the fact that numerous international studies have shown that civil SET is a vital component of economic and social progress.[57] The patterns of economic development of the most successful industrializing nations show that the benefits of innovation have not been produced by those nations which have prioritized military over civilian R&D.[58] This is partly because the secretiveness

[55] South African Department of Arts, Culture, Science and Technology, *South Africa's Green Paper on Science and Technology: Preparing for the 21st Century* (Government Printer: Pretoria, 1996).

[56] For a discussion on national innovation systems, see Nelson, R., *National Innovation Systems* (Oxford University Press: New York, 1993).

[57] See, e.g., Rosenberg, N., Landau, R. and Mowery, D., *Technology and the Wealth of Nations* (Stanford University Press: Stanford, Calif., 1992).

[58] Alic, J. *et al.*, *Beyond Spin-off: Military and Commercial Technologies in a Changing World* (Harvard University Press: Cambridge, Mass., 1992).

of the military sector and its institutional culture do not encourage a broad diffusion of commercial technologies. Moreover, the rapid evolution of global corporate networks over the past two decades has created a global economy where competitiveness is determined by the speed and efficiency with which civil innovations are brought to the market-place. This process of global techno-networking generates R&D investment on a scale which dwarfs the amounts that national defence budgets can mobilize. This is one of the major reasons why military R&D is no longer at the leading edge of technological innovation. Failure to adapt to such challenges can only mean further loss of competitiveness and a spiral of economic decline.

In South Africa the arms industry lobby persists in arguing that the defence sector provides a major contribution to South Africa's technological and hence competitive performance. Evidence to support this claim is sadly lacking.[59] During the apartheid era South Africa's R&D policy was 'mission-oriented', having been largely driven by military imperatives.[60] While military R&D effort was at the heart of the national system of innovation, military R&D expenditure was separated from civil.[61] In 1987, according to the Green Paper, 1.1 billion rand or 54 per cent of the total R&D budget was spent on defence. This fell to 283 million rand or 14 per cent of total R&D expenditure in 1993.[62] Since 1990 there has been a SET policy vacuum, following the erosion of the mission-oriented approach to R&D which was epitomized by the military and nuclear priorities of the apartheid state.[63]

In response to this policy vacuum and the generally poor performance of South Africa in innovation, the Green Paper argues for a deep and pervasive transformation of the existing SET base involving a high degree of cooperation between the government and civil society in the reformulation of R&D policy and resource allocation:

A new policy framework will have to take into account the distortions, inequities and failures of the existing SET system. The policy should include a commitment to reassure all South Africans that the future activities and benefits of the system will support the development agenda set out in the RDP. The values of race and gender equity, redress, efficiency, effectiveness, excellence, relevance and democracy will form the foundation of the new system. It should be characterised by greater accessibility to the public from the point of view of both access to information and ability to

---

[59] Kaplan, D., *Ensuring Technological Advance in South African Manufacturing Industry: Some Policy Proposals*, in Joffe *et al* (note 52), pp. 237–63.

[60] *Green Paper on Science and Technology* (note 55), p. 9.

[61] The Organisation for Economic Co-operation and Development (OECD) defines a national system of innovation as 'a network of institutions in the public and private sectors whose activities and actions initiate, import, modify and diffuse new technologies'. An alternative, fuller definition is 'a system of interacting private and public firms (either large or small), universities and government agencies aiming at the production of science and technology within national borders. Interaction among these units may be technical, commercial, legal, social and financial, in as much as the goal of interaction is the development, protection, financing or regulation of new science and technology'. Green Paper on Science and Technology (note 55), p. 25.

[62] The latest available figures. *SA Science and Technology Indicators* (Foundation for Research and Development: Pretoria, 1993).

[63] *Green Paper on Science and Technology* (note 55), p. 9.

contribute to decision making. The goals of the RDP represent the central context within which to implement a new SET framework for South Africa, being meeting basic needs, developing human resources, building the economy and democratising the state and society.[64]

In the future R&D policies will be oriented towards 'maximising and redirecting benefits to the population at large instead of to a minority'. In its statement on future policy goals and objectives the Green Paper argues that in developing a well-functioning national system of innovation there is a need to redirect 'expenditure and SET production capacity from defence to civil/commercial needs'.

As a small country with limited technological resources, South Africa needs to realign its R&D expenditure with its broader development goals. This requires a more coordinated approach to military and civilian R&D. At present military R&D is highly segregated from civil efforts at innovation and this means inefficiencies and duplication. Given the international trend for civil innovation to become not only the driver of competitive advantage but also an increasingly important source of generic technology for defence systems, the traditional wall of separation between civil and military R&D needs to be broken down. A more integrated approach to innovation in South Africa is urgently required, but one that is driven by commercial and 'national needs' policies.

Assuming that some resources will continue to be allocated to military research and development, the Green Paper recognizes that a new criterion needs to be established for a military R&D policy. This involves a critical look at how military R&D expenditure is allocated in terms of the returns to the economy. The Rooivalk programme, for instance, has had the lion's share of the military R&D budget, yet the benefits, assuming that it continues and a joint-venture partner is found, are likely to be to the management of Atlas and the workers of a foreign country (Malaysia) rather than to the South African labour force. Questions arise as to whether this is acceptable in the light of South Africa's employment situation. The implication of the situation is that the South African Government would be subsidizing R&D to the benefit of other nations. Equally, with the Olifant tank upgrade programme the question should be asked whether product development should continue in the light of the changing strategic doctrine and in view of the pressing social needs of the vast majority of South African people. These types of programme illustrate the continuing distortions in national R&D strategies that have been inherited from the apartheid system and highlight the urgent need to restructure the military R&D system in the light of the changing priorities of South African society.

The later White Paper on Science and Technology, while not as detailed or analytical as the Green Paper (which was specifically designed to stimulate a national debate on national innovation policy), reinforces the government's commitment to economic development and national needs through its science

<hr>

[64] *Green Paper on Science and Technology* (note 55), p. 15.

and technology policy: 'The development and application of science and technology within a national system of innovation (NSI) in South Africa will be central to the success of the Growth and Development Strategy (GDS) of the Government as it seeks to address the needs of all South Africans'. It goes on to stress that 'the development of innovative ideas, products, institutional arrangements and processes will enable the country to address more effectively the needs and aspirations of its citizens. This is particularly important within the context of the demands of global economic competitiveness, sustainable development and equity considerations related to the legacies of our past system'. While the discussion on military R&D is notably restrained in comparison to that of the Green Paper, it does stress the need to convert some of the skills tied up in military R&D activities to civilian innovative efforts: 'the defence sector in general is a repository of considerable skills in instrumentation, control and advanced materials handling. Extending or converting these skills to civil use could broaden our industrial skills base considerably'. The White Paper also stresses the importance of refocusing R&D policy away from the 'classical model of defence research, where research institutes exist to fulfill roles rather than follow the market' towards a dual-use model of innovation. However, it contests the idea of military R&D being phased out altogether in favour of civilian technology, arguing that this is untenable. Rather it reasons that 'the arms industry must make special efforts to leverage spin-offs in the civilian sector and to develop relationships with civilian institutions in the NSI to promote spin-ons. It is via this partnership route that our arms industry will achieve its rightful place in the mind of the public, and shake off negative perceptions, associated with its role in the South Africa of the past'.[65]

The difference in tone between the Green Paper and the White Paper reflects the compromises being made between the old order and the new at all policy levels in South African society. This compromise may fall short of the aspiration to convert the arms industry to civil production, advocated by those intent on full demilitarization. Nevertheless it has gone a long way towards reorienting military R&D policy by subordinating it to the broader socio-economic goals of South Africa's transition. Moreover, by integrating military R&D into the national system of innovation, the sector is being forced into greater transparency and accountabilty.

## VI. Conclusions

The de Klerk Government's 'hands off' policy towards the defence industrial base during the transition phase added to the costs of disarmament and failed to deliver any tangible economic benefits in the form of a peace dividend. The decline in defence expenditure and the exposure of the arms industry to market forces have resulted in a general decline in the performance of defence con-

[65] South African Department of Arts, Culture, Science and Technology, *White Paper on Science and Technology: Preparing for the 21st Century* (Government Printer: Pretoria, Sep. 1996), pp. 5, 41, 42.

tractors and have led the survivors to adopt offensive strategies based on increased concentration and monopolization of the defence market, rather than diversification or conversion to civil production.

The new government inherited a smaller arms industry facing enormous adjustment problems. While support for the use of the peace dividend to convert the defence industrial base to civilian production has grown, there has been little evidence of successful conversion in South Africa, partly because of the formidable barriers to conversion at the level of defence firms, but also because the new government has been slow to develop a coherent defence industrial strategy that addresses conversion issues. Instead the Government of National Unity chose to reallocate budgetary resources to RDP goals, leaving defence industrial adjustment to market forces. It is questionable whether this has produced the optimal economic outcome.

By the time the new government has formulated its defence industrial policy, it may well be too late to instigate a national conversion strategy (assuming that one is adopted) because the arms industry in its bid for survival has already diversified into those civilian markets that are available, down-sized those operations that were unprofitable and moved into international markets through joint ventures and strategic alliances where possible. Thus the window of opportunity to use the conversion of the arms industry as a means of restructuring and revitalizing the country's industrial base may have been lost through the inevitable delays of the transition process.

# 9. Summary and conclusions

The purpose of this book is twofold: to provide a comprehensive empirical analysis of the wide-ranging adjustments that have taken place within South Africa's arms industry during the country's transition to democracy; and at a more general level to locate this process of defence industrial adjustment within the broader context of South Africa's changing political, strategic and economic environment. It also attempts to place the South African experience in the context of broader debates about the economic effects of military expenditure and defence industrialization and the relationship between disarmament and development in developing countries. This chapter summarizes the findings and outlines those issues of defence industrial policy that remain unresolved.

## I. Summary

The structure of this book reflects the distinct periods of defence industrialization demarcated by three distinct phases of South Africa's recent history—militarization, disarmament and demilitarization. This methodology, together with a review of conceptual issues relating to the economic effects of military expenditure and defence industrialization and of the implications of the reformulation of security concepts in the light of the changing regional and international geo-stategic environment, informs the analysis. These issues are discussed in chapter 1.

Chapter 2 describes the evolution and development of South Africa's arms industry during the militarization period between 1975 and 1989. During that period South Africa experienced a form of 'total war' which involved the militarization of society and the increasing use of state violence in defence of apartheid. This process of militarization operated at the political, economic and ideological levels, affected all levels of society and encroached upon the lives of all South Africans. At an economic level the process of militarization involved significant increases in military expenditure, the development of an arms industry, and growing institutional links between the state, the military and private industry, resulting in the emergence of a military–industrial complex conditioned by the features of apartheid. The development and expansion of South Africa's arms industry after 1975, which was prompted by the imposition of the UN arms embargo in 1977 and the country's increasing military involvement in a number of regional conflicts, thus took place within the context of the broader process of militarization.

By the end of the 1980s South Africa's arms industry had developed into one of the most significant sectors in the country's industrial base in terms of employment, output and the number of firms involved. The economic significance of the arms industry was a function of its privileged position in terms of access to state resources. During the 1970s and 1980s the state invested heavily in the arms industry at the expense of other sectors of the industrial base. This 'misinvestment', in which strategic industries like the arms industry absorbed a disproportionate share of scarce resources (capital, labour and technology) distorted the country's industrial development and was a contributing factor to the deteriorating performance of the economy during the late 1970s and throughout the 1980s. The fact that the development of South Africa's arms industry between 1975 and 1989 contributed to the economy's declining performance, by crowding out civil innovation and absorbing scarce investment resources, provides a valuable contribution to the international debate on the economic effects of military expenditure and defence industrialization highlighted in chapter 1.

Chapter 3 examines South Africa's changing strategic, political and economic environment during the transition to democracy between 1989 and April 1994. It identifies the factors operating at both internal and external levels which prompted the de Klerk Government and the ANC to pursue a negotiated end to apartheid. Domestic factors included growing levels of internal political violence and opposition to apartheid, together with the failure of Total Strategy to co-opt a significant proportion of black South Africans into some kind of power sharing with the apartheid government. External factors included trade and financial sanctions, the end of the cold war and the demise of superpower rivalry in Southern Africa, and the SADF's loss of air superiority in southern Angola in 1988.

South Africa's transition to democracy, which started in 1989 and culminated with the holding of its first democratic elections in April 1994, was accompanied by dramatic changes to its strategic environment. The ending of apartheid and South Africa's military withdrawal from the Southern African region removed the dominant source of instability in the region and led to a dramatic improvement in interstate relations in the region. This in turn prompted moves towards greater regional cooperation on defence and security matters within the context of the SADC.

South Africa's domestic political setting also witnessed dramatic changes between 1989 and 1994. After 1989 de Klerk initiated a number of fundamental political reforms which included the unbanning of opposition movements and the repeal of apartheid legislation. Dominant features of de Klerk's reforms were the re-establishment of civilian control over the military and the liberalization of certain aspects of state security policy. Despite de Klerk's attempts to reduce the power and influence of the military within state decision making, he was unwilling to effect a full transformation of the apartheid military establishment. Thus the institutional and ideological manifestations of militarism from the apartheid era remained largely intact during the transition period.

In terms of South Africa's economic situation, between 1989 and 1994 the domestic economy experienced its longest and worst recession since the great depression of the 1930s. The recession, which had its origins in the 1980s as a result of sanctions, disinvestment and the increasing structural problems associated with maintaining the apartheid economy, was also influenced by the global recession (particularly among South Africa's major trading partners), the government's inability to control expenditure, which aggravated inflation, a severe drought in 1992–93 and a lack of investor confidence because of the uncertainty surrounding South Africa's political future. The severe deterioration in South Africa's economic situation after 1989 certainly played a role in prompting de Klerk to initiate his political reforms. However, it also at some points undermined the positive progress that was being achieved on the political front by exacerbating the existing socio-economic problems inherited from the apartheid era.

Chapter 4 examines the nature and impact of the defence cuts and disarmament measures which were implemented in South Africa between 1989 and 1994. It highlights the facts that the cuts in defence expenditure were implemented in the context of a severe domestic economic recession and that they probably contributed to the deteriorating performance of the economy during the same period. This chapter also shows how the defence cuts had a significant impact on the size, structure and performance of the domestic arms industry at an aggregate level and how this was reflected in the industry's declining sales and output and large-scale retrenchments.

Chapter 5 considers the adjustment experiences of the South African arms industry between 1989 and 1994 at the micro-economic level. It examines the various supply-side adjustment strategies which public- and private-sector defence firms pursued in response to the defence cuts and declining domestic demand for armaments. From the information presented in this chapter it is evident that most public- and private-sector defence firms pursued a combination of defensive and offensive adjustment strategies. The most popular defensive strategies, which were aimed at reducing firms' defence business, involved retrenching workers, down-sizing defence operations and diversification. The most popular offensive adjustment strategies, which were aimed at increasing firms' defence business, included increasing exports, concentration and monopolization of (local and foreign) defence markets, and international collaboration. The wide range of adjustment strategies reflected the diverse nature of the country's defence industrial base. It also reflected the absence of any coherent government strategy for helping firms to adjust to lower levels of defence expenditure. As a result of the lack of any government policy towards the industry, a significant amount of valuable skills and technologies, which were embodied in the arms industry, were either lost or wasted. When the ANC-led government came to power in April 1994 it inherited a domestic arms industry which was substantially smaller, more concentrated and less cohesive than it had been during the apartheid era. Even so, in April 1994 the apartheid-era military–industrial complex still remained largely intact.

With the demise of apartheid and the withdrawal of the superpowers from the Southern African region, the new ANC-led government has been attempting to redefine its relations both with the region and with the rest of the world. At the same time its first-ever non-racial democratic elections have triggered a profound normative, institutional and cultural transformation of South African society. In chapter 6 these issues are examined in the light of their effect on the security community. In particular the chapter examines the recasting of the security agenda, forms of regional cooperation, including peacekeeping, arms sales in the region, the military backlash, and strategic doctrine and its implications for the arms industry.

Although the new government faces a more benign regional security environment it has found itself confronted with a new set of challenges which threaten the fragile peace and stability of the region. These include migration, refugees, small-arms proliferation, landmines, environmental degradation, food insecurity, poverty and violent crime, many of which are legacies from the apartheid era. These new threats to regional stability have demanded new responses to security provision. The government has adopted a broadly based development approach to regional security which has been given formal policy recognition within the Reconstruction and Development Programme and the 1996 White Paper on National Defence. The adoption of a more holistic approach to security reflects a profound rejection of the narrow militaristic notion of security from the apartheid era.

The fact that the new security threats transcend national borders has encouraged an emphasis on collective solutions, which has resulted in attempts to build a common security regime in the Southern African region through the SADC. Such initiatives are at a very early stage of development and there are many pitfalls to be overcome, not least the residual fears of South Africa's neighbours about its overwhelming economic and military power.

To date the most advanced form of regional cooperation has been military collaboration, particularly in the attempts to develop a regional peacekeeping force under the guidance of the BMATT team. The emphasis on military forms of cooperation has evolved partly as a result of the SANDF's search for new roles and functions in the post-apartheid era. Linked to the trend in military cooperation has been the promotion of arms sales within the Southern African region in an attempt to find new markets for South Africa's ailing arms industry. These trends reflect the military establishment's attempts to assert itself in the face of concerted efforts to demilitarize South Africa's security agenda. Nowhere is this more apparent than in the debate on strategic doctrine, in which the military has made some progress towards reversing the decision to adopt a defensive defence posture. That these trends appear to contradict the stress placed on non-military forms of security cooperation in both SADC documents and the White Paper on National Defence reflects the compromise and accommodation that have occurred between the old order and the new, a dominant feature of the present transition process and a symptom of South Africa's 'negotiated revolution'.

The outcome of the debate on strategic doctrine which has taken place through the Defence Review will inform future procurement choices and will thus have implications for the defence industrial base. Even if an offensive posture is retained there is no longer a guarantee that weapon systems acquired by the SANDF will be purchased from domestic sources. Indeed, the proposed purchase of corvettes, combat aircraft and submarines from the UK suggests that given the choice of off-the-shelf products in the context of budgetary constraints the SANDF's loyalties to the domestic defence industrial base are on the decline.

The current emphasis on transparency and accountability reflected in the new government's commitment to reasserting civilian control over military institutions is captured in chapter 7, which deals with the institutional and political changes in the security establishment since 1994. The re-balancing of civil–military relations in favour of civilian authority is reflected in the establishment of a new civilian Defence Secretariat and in the functioning of a powerful parliamentary Defence Select Committee. The more transparent and civilian-oriented approach to military matters is reflected in the open, consultative nature of the White Paper on National Defence and the Defence Review during 1997. These measures are an essential element of the demilitarization of institutions, structures and culture that were built up by the military during the apartheid era. In effect the reforms have distanced the military–industrial complex from the executive levels of the state, reducing its power and influence and leading to a fragmentation of interests between the state, the military and private capital and to a dissipation of the military–industrial complex.

Moreover, the increasing prominence of civilians, including parliamentarians, in the defence decision-making process has had a significant impact on the defence industrial base. The arms industry has come under much closer public scrutiny than was the case during the apartheid era. As a result the industry has been forced to become more responsive to public opinion and has thus become an enthusiastic supporter of the RDP.

In the context of continuing defence cuts in order to fund the RDP and in the absence of a government policy on the defence industrial base, the arms industry has been forced to down-size even further. Adjustment strategies have focused on exports and joint ventures, particularly since the lifting of the UN arms embargo. The latter has meant, however, that the industry is subject to greater competition from foreign companies who are now tendering for SANDF contracts. By 1997 the arms industry was significantly smaller, more concentrated and struggling for survival. In 1997 the government finally initiated a White Paper on the Defence Industry, with a remit to develop a defence industrial adjustment strategy, but by the time the review process is completed in 1998 there may be little left to adjust.

In a country which has inherited huge inequalities in income and economic power, one of the major challenges facing the new government has been to redistribute resources. This it has tried to do by redirecting the 'peace dividend' to the RDP. This is a form of macroeconomic conversion which puts socio-economic development at the heart of the strategy. This and other conversion issues are dealt with in chapter 8, which examines the national debate on conversion and the peace dividend. The socio-economic costs of the down-sizing of South Africa's defence industrial base have generated considerable public support for a national conversion policy. While the government has addressed the macroeconomic conversion issues it has done little in the way of addressing micro-economic conversion at the level of individual defence firms or defence-dependent communities. Pressure to use the peace dividend to convert the defence industrial base to civilian production has grown, but so far the evidence of successful micro-conversion in South Africa has been scanty. This is partly because of the barriers to exit from the defence market, but also because the new government has been slow to develop a coherent defence industrial strategy that might address micro-economic conversion issues. This will, however, be addressed in the White Paper on the Defence Industry.

Despite the much reduced size of the arms industry there is still an urgent need for a coherent defence industrial policy. A number of key policy issues, such as procurement, R&D, privatization, arms trade and conversion, require clearer policy guidelines.

## II. Defence industrial policy issues

Any future policy on the arms industry should address a number of key issues.

First, given the central role of government in creating and maintaining the South African arms industry during the apartheid era, it is understandable that questions should arise about its future role towards the arms industry in the post-apartheid era. As the single largest buyer from the arms industry, the South African Government has enormous influence over the size, structure, strategies, output, profitability, entry and exit, exports, efficiency and ownership of the industry. Through its procurement function the government has created a highly protected and regulated market which differs significantly from normal commercial markets. Thus the issue is not whether the government should have a role in defence industrial policy, but what form this role should take.

One view is that defence procurement should be used as an instrument of national industrial policy, awarding contracts to support key technologies and to back national 'winners', in effect a military Keynesian policy of industrial intervention. This approach represents a continuation of past practices, where the arms industry is classified as a strategic industry with privileged access to state subsidies and expenditure. However, the future is viewed as uncertain and unpredictable: today's winners might be tomorrow's losers. A 'national champions' approach could be a high-risk strategy, particularly in view of the fact that the government has little experience as an entrepreneur. A more laissez-faire approach involves leaving the size and structure of the arms industry to be determined by market forces. This 'hands-off' approach to the sector stresses the importance of securing value for money from the procurement budget through competition policy.

Beyond simple defence industrial concerns it is sometimes argued that defence policy should be used in pursuit of the more general economic goals of industrialization and innovation. It is perhaps axiomatic that if the government is to adopt interventionist industrial, employment and technology policies the responsibility is best left to the appropriate civilian ministries such as the Department of Trade and Industry. Moreover, using defence procurement as an instrument of industrial policy raises fundamental questions about the objectives of defence policy itself. There is strong evidence to suggest that where broader economic concerns feed into the procurement process a procurement agency becomes hostage to the narrow vested interests of the arms industry, which may not help wider economic and social goals. Furthermore, where economic rather than strategic priorities dominate procurement decisions, this can have a negative impact on the type of equipment which is procured for the armed forces.

Given the very real fiscal constraints facing the South African Government, a careful evaluation of the costs and benefits of alternative defence industrial policy scenarios needs to be undertaken, paying particular attention to both their strategic and the economic implications. In the process of policy formulation emphasis should be given to an analysis that is independent of the vested interests of the arms industry and closely aligned institutions in order to guarantee the greatest possible objectivity when evaluating outcomes.

Second, a primary consideration in formulating policy on the arms industry is to identify the desired characteristics of a post-apartheid defence industrial base. These will be determined partly by the country's strategic objectives, particularly in the light of the changed strategic environment. A related issue is whether there are lower-cost options for meeting these strategic requirements than those employed at present, for example, through inventories and stockpiles, licensed production or procurement of components from civil industry (such as semiconductors and computers) and via imports. The Ministry of Defence and the SANDF are keen to retain strategic industrial assets in certain sectors such as communication systems and electronic warfare and the critical skills and technologies required for maintaining and upgrading weapon systems. A related concern is the need to guarantee the supply of essential *matériel* such as ammunition and spare parts during periods of conflict.

## Procurement issues

A major policy issue which requires urgent consideration is which institution or government department should be responsible for procurement. At present Armscor has responsibility for armaments procurement, but in the light of the revelations made by the Cameron Commission about Armscor's illegal trading it is clear that the organization lacks the legitimacy and accountability which are needed in South Africa's new political dispensation. Armscor has also developed its own corporate vested interests which have often appeared at odds with the country's political and economic realities. It is suggested that the procurement function could be relocated in the civilian-controlled Defence Secretariat in order to ensure greater accountability and transparency. Armscor personnel associated with technical functions such as quality control and testing and evaluation could be transferred to the Secretariat.

A second issue concerning procurement is whether parliament should be involved in the procurement decision-making process. Parliamentary involvement can take one of two forms—proactive or reactive. In the USA, for instance, Congress is involved in approving major procurement projects. In the UK all decisions are made by cabinet committee and only after the fact are they reported to the House of Commons. The British Select Committee on Defence has a remit to review decisions and comment on them, but has only indirect influence and no power of veto over defence procurement decisions. In a situation in which the new ANC Government is intent on asserting greater accountability and transparency in defence decision making, the more proactive model would seem more appropriate to South Africa's new political needs.

In the interests of greater transparency with respect to procurement decision making, Armscor or the equivalent institution responsible for procurement should at the very least present an annual report to parliament providing details of all major current and proposed procurement projects, including financial details. This measure, which is in place in many countries, is designed to help monitor cost controls and delays on major procurement programmes in order to prevent 'gold-plating' and other abuses of the defence contracting system.

A third and major problem facing procurement policy is the issue of cost controls. Weapon systems, particularly high-technology equipment, are costly and the trends are towards rising costs in real terms. Each generation of equipment is significantly costlier than the previous generation which it replaces. Typically the unit costs of major weapon platforms rise by about 10 per cent per annum in real terms, and since the procurement budget cannot hope to match such price increases there are inevitable downward pressures on the number of weapons procured with an associated need to restructure the arms industry and force structures. In the face of such unavoidable trends, the South African Government needs to to be vigilant over the efficacy of its procurement policies if it is to keep costs under control.

There is a range of cost-controlling mechanisms the South African Ministry of Defence may want to consider:

1. Opening up the defence market. This is one way to increase competition in order to prevent the rent-seeking behaviour associated with monopoly or oligopoly in the defence market.

2. Regulations concerning profits on defence contracts. These can be tightened up through the use of fixed-price contracts as opposed to cost-plus contracts in order to help prevent gold-plating and excessive profits.

3. Cost controls on the life-cycle costs of weapon systems. Life-cycle cost estimates should be brought into the public domain for scrutiny, but attention should also be paid to reducing life-cycle costs through the product design process, that is, through the introduction of modularity in product design which has

been found to make significant life-cycle cost savings in other arms-producing countries such as the UK.

4. The assessment of each major procurement requirement in the light of the range of possible solutions for the desired outcome. For example, if an offensive air defence system is required there is a choice to be made between combat aircraft and air defence missiles. The cost implications of these equipment choices can be very significant. Combat aircraft cost far more than missiles, the life-cycle costs of each aircraft often match the initial purchase price, and the costs of fighter pilots are considerable.

5. Cost-effectiveness studies. These are another way in which the costs of procurement programmes can be assessed, and they should be undertaken for all proposed programmes and made available for scrutiny by the parliamentary Defence Select Committee.

6. More attention to trade-offs within the procurement budgeting process. Costs need to be assessed in relation to the security benefits, outputs or effectiveness. This helps focus attention on the efficacy of defence procurement choices and prevents traditional inter-service rivalry or the monopoly of one service over others in influencing procurement decisions.

## Privatization

In the light of government policy with respect to the privatization and/or restructuring of state assets, it is important to consider whether Denel or parts of it should be privatized. The general issue of privatization has become a major controversy in South African society, provoking widespread opposition from certain groups such as COSATU. A similar response might be expected from Denel workers if privatization were to proceed. Privatization is very likely to involve further rationalization of Denel's workforce as the company tries to become more commercially viable. However, the possibility of opposition should not influence rational choices concerning privatization if this is the course of action chosen. The possible privatization of Denel would also fundamentally alter the company's 'cosy relationship' with Armscor and its dominance of the domestic defence market.

Evidence from other countries such as the UK suggests that privatization may result in improved company performance measured in terms of labour and total factor productivity. These improvements should in time feed through to affect the unit costs and prices of defence equipment. Concern has been expressed, however, that the UK's privatization and competition policy has resulted in a greater concentration of the arms industry, resulting in the formation of national monopolies which are associated with rent-seeking behaviour. While there are strong arguments in favour of privatization, some of Denel's facilities are not commercially viable (for instance, the Overberg Testing Range) but are nevertheless strategically important for the armed forces. The state may therefore wish to retain these strategic facilities within the public sector, or it may consider allowing them to be managed by the private sector while retaining ownership itself. Decisions about the privatization of Denel or parts of it should be decided on a case-by-case basis, with all economic and strategic variables being taken into consideration.

## R&D and technology policy

The South African Government is in the process of restructuring its R&D priorities. Traditionally large amounts of state resources were invested in military R&D, with few benefits to civil society or the national technological base. The changing strategic doctrine implies a shift in procurement from offensive to more defensive systems, which will promote a restructuring of military R&D priorities. The issue of restructuring the military innovation system is inevitably linked to the question of national technological priorities and whether or not investment in military technology is beneficial to the civilian economy. The government's new science and technology policy, encapsulated in the 1996 White Paper on Science and Technology, which updated the 1996 Green Paper, has highlighted the need to reorient South Africa's R&D policy away from defence-related goals to goals that emphasize socio-economic improvement. Many nations facing reduced budgets for military R&D and a changing dynamic in the relations between civil and military technology have shifted their R&D policies towards strategies which maximize dual use, and this has required a fundamental change in R&D institutional structures and policy priorities. So far South Africa has been slow in addressing the prioritization of dual-use technology; however, it can be assumed that the inexorable pressures of globalization will force its policy makers to adapt their R&D policies to such goals in the near future.

If military R&D is integrated into a National System of Innovation, as proposed in the 1996 White Paper on Science and Technology, this could have significant implications for the future structure of the South African arms industry. For example, a national policy which targets national needs and export competitiveness is likely to place emphasis on process technologies rather than the product technologies which are characteristic of military R&D. This may provide diversification opportunities for certain defence sectors such as electronics and mechanical engineering but would prove inimical to the interests of others such as large system integrators.

## Conversion and diversification

The processes of disarmament and demilitarization which began in 1989 are crucial elements in the dismantling of the apartheid system. The positive aspects have, however, a negative corollary in the displacement of labour, loss of skills and decline in industrial output which have accompanied the downsizing of the arms industry. For the ANC Government the challenge it faces here is to optimize the transfer of resources from the military to the civilian sector, via conversion or diversification, with the minimum of dislocation.

In the absence of any government policy towards the arms industry since 1989, defence firms have been expected to adjust to the cuts in defence expenditure in accordance with market forces. Strategies of diversification and conversion which have been pursued by both public- and private-sector companies have met with only limited success. More often than not defence firms have focused on producing civilian prototypes of military products without doing any market research to assess whether the demand existed for such products, reflecting a limited knowledge of how commercial markets work. The failure of conversion efforts at micro-economic level is not unique to South Africa. The international evidence suggests that only those companies with relative strengths in dual-use technology markets, or those at the lower levels of the product hierarchy (producing components and sub-systems), have found the adjustment to civilian markets less problematic. So far there have been no attempts in South Africa to reproduce the classic plant-based form of conversion as suggested by the Lucas Corporate Plan in the UK in the 1970s. This is not surprising given the tremendous barriers to exit from the defence market and the severe domestic recession which accompanied the decline in the domestic demand for armaments.

The failure of the market to reallocate resources from defence to the civil sector has raised the issue whether the government should take a more interventionist role in the processes of conversion and diversification. In South Africa support for a government-orchestrated conversion strategy has come from a number of quarters, including certain trade unions, church groups and elements within the ANC. However, apart from a rhetorical call of 'swords into ploughshares' before the elections, the new ANC Government has done little to formulate a policy on the arms industry that seeks to optimize the gains from demilitarization. Understandably in the absence of a coherent defence industrial adjustment policy, defence companies have continued to pursue strategies aimed at increasing their share of both local and foreign defence markets.

If the South African Government is to adopt a policy of conversion following the completion of its White Paper on the Defence Industry, it needs to think carefully about the model of conversion it might adopt. Internationally a number of models have been tried and tested with varying degrees of success and failure. The 'top-down' approach implied in government-orchestrated conversion strategies has not been particularly successful in Eastern Europe and the former Soviet Union. There has also been relatively little success with the micro-conversion initiatives that were popular in the 1970s and 1980s in Western Europe. The lack of success with these different approaches to conversion suggests that an alternative approach to defence industrial restructuring needs to be identified for South Africa's arms industry.

Political conversion, an approach which has emerged as a result of past conversion failures, involves an eclectic use of policy mechanisms at the national, regional and local levels to encourage defence-dependent communities to diversify their local economies away from defence production while utilizing the skills of displaced defence workers. This model of conversion is in essence concerned with the demilitarization of economic activities, while capturing the long-run economic benefits of the peace dividend. In this context the role of government is to promote a macroeconomic environment conducive to the conversion of resources from military to civilian use. This is achieved through mechanisms and incentives, such as tax breaks, subsidies to spin-off companies, retraining programmes for defence workers, and economic regeneration programmes for regions or localities affected by the defence downturn. These policies are being implemented in a number of regions in Europe and the USA.

A key issue which distinguishes the South African experience from other attempts at conversion is the racial characteristics of the arms industry. The apartheid arms industry was consciously structured as an enclave to create jobs for Afrikaners and to promote the interests of Afrikaner capital. Any conversion policy must therefore be sensitive to the need to abolish this legacy of apartheid by redirecting resources away from the defence sector and towards those areas of the economy that will have maximum benefits for the people who have been disadvantaged in the past. In this sense conversion is an important transition mechanism designed explicitly to demilitarize those components of the South African industrial base that were part of the apartheid military–industrial complex and which have survived into the post-apartheid era.

## Internationalization of armaments production

The increasing costs of national development programmes and the small production runs for the domestic market have provided an economic incentive for the South African arms industry to internationalize its operations. Since the end of the cold war arms industries all over the world have been forced to become simultaneously more internationalized and more competitive in order to survive. If the South African arms industry wishes to survive and remain viable then it has no other choice than to follow suit. Internationalization takes a variety of forms: collaboration, licensed production, co-production and offsets.

In theory international collaboration with one or more partners provides cost savings for both R&D and production. There can also be industrial benefits in terms of technology transfers and larger economies of scale. However, in practice there may be significant costs. Bargaining between governments and bureaucracies can lead to inefficiencies through time-delays and administrative costs. Work is often allocated on the basis of equity inputs rather than efficiency and comparative advantage. Collaboration also involves high transaction costs reflected in the duplication of management structures and decision-making processes. The net effect is delays, cost overruns and high unit costs.

Co-production is a form of collaboration on production only and involves one or more partners in a fully integrated production programme. Such programmes are less bureaucratic than collaborative programmes but may involve penalties reflected in the costs of market entry, the costs of transferring technology and the costs of learning new technology. However, the benefits of co-production (technology transfer, employment generation, import savings and industrialization) are generally thought to outweigh the costs.

Licensed production is a traditional form of direct offset by which the purchasing nation builds foreign-designed equipment under licence in its own country. In a buyer's defence market, more and more customers insist on licensed production rather than off-the-shelf purchases. While South Africa has benefited from licensed production in the past, which helps to facilitate technology transfer from supplier to buyer, as a supplier of defence equipment it is likely to find itself having to agree to licensed production agreements. For example, if Denel manages to secure the Rooivalk helicopter contract with Malaysia, which seems increasingly likely, it will probably involve some licensed production in Malaysia. While the sale will increase Denel's profits it will not enhance the employment of South African defence workers as production will in all likelihood take place in Malaysia.

Offsets are a growing feature of the international arms trade and a major mechanism through which countries such as South Africa can acquire technological know-how. Offsets enable the purchaser to recover some or all of the purchase costs through the relocation of economic activity to the country of purchase. Direct offsets may include the production of spares and parts for the weapon system purchased or there may be indirect offsets for some non-related economic activity. Maximizing offsets has become one of South Africa's biggest challenges now that it is formally reintegrated into the international community. Currently all contracts involving South African arms purchases from foreign sources of supply must have a minimum of 60 per cent in offsets of the total value of the purchase. There are, however, concerns about the extent to which offsets represent genuinely new business which would not have been obtained without the offset agreement. Offsets can also have a negative impact on local industry if they involve dumping from the supplier country, or if they replace domestic demand. While South Africa may benefit from offsets, as a seller it can also expect increasing pressures to provide offsets to foreign customers. This many involve the risk of it losing some of its competitive advantages in the long term as the transfer of technological know-how fosters potential competition.

South Africa's defence firms are establishing a number of joint ventures with foreign defence companies. A major policy question concerning this trend is whether the government should attempt to regulate these forms of internationalization. As most forms of internationalization involve technology transfer there is the question of international norms on military technology transfers. Should South Africa, for instance, involve itself in joint ventures with countries such as Indonesia and Taiwan given their dubious international standing at present? If such decisions are subject to regulation, should they be subject to parliamentary scrutiny in the same spirit of transparency as is applied to the arms trade?

## Arms trade issues

In the light of the two reports of the Cameron Commission and the establishment of the NCACC, new policies and decision-making procedures for South
Africa's arms trade and parliament have been formulated. These mechanisms
and procedures are guided by a new set of norms and values designed to cast
off South Africa's reputation as a pariah state and to bring it into line with internationally acceptable principles and practices.

Notwithstanding the implementation of new arms trade policies and decision-
making procedures (such as the NCACC), there are still a number of key policy
issues with respect to South Africa's arms trade which have not been resolved,
and which were highlighted in the Cameron Commission's second report.

First, control over the export and import of different types of armaments has
not been centralized in a single government agency. The NCACC, Department
of Trade and Industry and Department of Safety and Security are responsible,
respectively, for the import and export of conventional armaments, weapons of
mass destruction and commercial small arms and ammunition. There is an
urgent need to centralize control in one government agency in order to avoid
policy inconsistencies and breakdowns in control.

Second, the government is still adamantly opposed to parliament playing any
decision-making or advisory role with respect to prospective arms transfers.
While it is willing to allow the parliamentary committees on defence and
foreign affairs to scrutinize arms transfers retrospectively, it is not interested in
allowing the legislature the power to 'second-guess' the executive with regard
to sensitive foreign-policy issues. The absence of any formal role for parliament
allows the executive considerable discretion in interpreting the NCACC's
criteria for approving arms sales. It also prevents any rigorous evaluation of the
executive's adherence to the NCACC's guidelines and undermines the ANC
Government's oft-stated commitment to transparency and accountability.

The relationship between human rights and arms transfers has become a particularly important issue with respect to South Africa's arms sales given the
centrality of human-rights issues to both the ANC and various groups within
civil society. Certain groups, particularly the churches and the peace movement,
have voiced strong moral objections to South Africa's continued involvement in
the international arms trade. In particular there is growing disquiet about its
attempts to promote its trade within sub-Saharan Africa given the continent's
development crisis and high levels of intra-state conflict. It can be expected that
the debate on the arms trade will continue for some time to come, particularly
as private individuals and companies from South Africa have been implicated
in illegal arms sales to countries such as Angola, Burundi, Rwanda and Sierre
Leone.

Advocates of the arms trade continue to argue that arms transfers are necessary to further South Africa's foreign and economic interests. However, there
has been a healthy questioning of these assumptions, reflected in the sub-

missions to the Cameron Commission and the public outcry about the proposed Syrian deal. In particular there has been a growing awareness that regional stability can be compromised by exports of arms to neighbouring countries that have few national resources. Moreover, arms exports to military and undemocratic regimes run counter to South Africa's stated policy of promoting democracy and human rights. The economic benefits have also been exaggerated, given the high level of subsidies that arms exports receive from the state. The government thus has some way to go before it finds an acceptable balance between responsibility, restraint and support for the arms industry's arms export drive.

## III. Concluding remarks

South Africa's historic transition to democracy has provided the new ANC-led government with a unique opportunity to develop new and innovative policies on defence and security matters. It has also offered it the chance to use the resources released from the cuts in defence expenditure—the peace dividend—to restructure and revitalize the country's industrial base and to eradicate the destructive legacy of militarism that is so pervasive in South and Southern Africa.

The reallocation of defence resources to civilian purposes as a result of defence cuts does not occur automatically. Rather, the relationship between disarmament and development needs to be politically constructed. The new government, through the articulation of the RDP, has explicitly created a link between disarmament and development and tried to develop formal macroeconomic mechanisms in order to redistribute resources from the military to development purposes. The RDP is thus a unique example of politically-motivated macroeconomic conversion.

Since April 1994 the government has initiated a number of public policy processes on defence and security matters. The public, transparent and highly consultative nature of these initiatives, which has provided much of the material and insights for this study, is a unique feature of policy making in post-apartheid South Africa and to the authors' knowledge has not been replicated elsewhere in the developing world.

Two debates have dominated these defence and security policy processes since April 1994. The first is concerned with the relationship between military expenditure and development and with the question how much military expenditure is enough in the context of post-apartheid spending priorities. The second has focused on reformulating South Africa's security and foreign policies in the light of new thinking on security and is concerned with the non-military aspects of security and sustainable peace and human development. These debates, which are linked to a common preoccupation with development and redistribution issues, have had a profound impact on the position and size of the arms industry which was built up during the apartheid era.

The lack of a clear government policy on the arms industry and the government's 'hands-off' approach to defence industrial adjustment since 1989 have resulted in the dramatic down-sizing of the domestic arms industry with the attendant loss of valuable skills and technologies. Up to this point the failure of government to implement a national conversion strategy as an integral part of a national industrial, science and technology policy has meant that a valuable opportunity to capture the benefits of disarmament and demilitarization may have been lost. While the start of a process to produce a White Paper on the Defence Industry is to be welcomed, the opportunity to maximize the benefits of defence industrial conversion may have passed.

Debates about the future of the arms industry are likely to be characterized by differing points of view. On the one hand critics of the industry will point to the drain on national resources and the opportunity costs of maintaining a domestic industrial base when there are pressing social needs to be met as a result of the legacies of apartheid and underdevelopment. On the other hand its supporters will argue that the arms industry provides a means of acquiring new technology and acts as a growth point for industrialization as well as contributing to import savings and valuable foreign currency earnings. These countervailing arguments on the costs and benefits of the national defence industrial base to the South African economy will remain central to debates on national spending priorities, particularly in the context of the RDP and the government's macro-economic policies.

In conclusion, the South African experience provides a clear example of the need to construct a link between disarmament and development through explicit government policies. It also provides useful insights for other developing countries who have sizeable defence sectors and who are experiencing a process of disarmament and demilitarization.

# Appendix 1. SADF rationalization and restructuring measures

## I. Army

The following army units/bases/installations were scaled down, disbanded or closed:

- Army Group Headquarters scaled down;
- special training project for SA Cape Corps (Coloured) volunteers in Kimberley terminated;
- horse breeding farm at De Aar closed;
- no. 4 Electronic Workshop in Youngsfield disbanded;
- no. 6 Signal Unit disbanded;
- Publication and Map Depot became a sub-unit of no. 1 Military Printing Unit in Pretoria;
- no. 91 Ammunition Depot in Pretoria and no. 92 Ammunition Depot in Witbank amalgamated into 'new' no. 91 Ammunition Depot in Naboomspruit; and
- conventional and counter-insurgency units of the part-time force rationalized into three conventional formations and three counter-insurgency formations.

## II. Air Force

The following air force units/bases/installations were scaled down, disbanded or closed:

- Air Force Base (AFB) Port Elizabeth closed;
- no. 16 Squadron (Alouette helicopters) disbanded;
- no. 12 Squadron (Canberras) at Waterkloof (Pretoria) disbanded;
- no. 25 Squadron (Dakotas) at Ysterplaat (Cape Town) disbanded;
- no. 27 Squadron (Albatrosses) at Ysterplaat (Cape Town) disbanded;
- Maritime Operational Conversion Unit at Ysterplaat disbanded;
- air commando squadrons 107 AFB Bloemspruit and 114 AFB Swartkop disbanded;
- Southern and Western Air Commands scaled down;
- AFB Potchefstroom closed;
- no. 10 Squadron (Cessnas) disbanded;
- no. 42 Squadron (Bosboks) disbanded;
- no. 84 Light Aircraft Flying School closed;
- no. 4 Squadron (Impala II) at Lanseria disbanded;
- no. 24 Squadron (Buccaneers) at Waterkloof disbanded;
- no. 30 Squadron (Puma/Super Frelon helicopters) at Ysterplaat disbanded;
- no. 81 Light Aircraft Flying School at AFB Swartkop (Lanseria) closed;
- Klippan Control and Reporting Post (Mafekeng) closed;
- Force Station Snake Valley (Pretoria) closed;
- no. 402 Aerodrome Maintenance Unit at AFB Ysterplaat (Cape Town) closed; and

– various aircraft—Canberras, Super Frelon helicopters, Westland Wasp helicopters, Kudus, Albatrosses, Harvards and Buccaneers—withdrawn from service.

## III. Navy

The following navy units/bases/installations were disbanded, scaled down or closed:

– marines disbanded;
– harbour protection units at Richards Bay, Durban, Port Elizabeth, East London and Cape Town closed;
– Naval Base Walvis Bay and Naval Base Simonstown scaled down;
– Naval Command West (Silvermine) and Naval Command East (Durban) disbanded; all naval operations shifted to Naval Headquarters (Pretoria);
– Naval Dockyard, Simonstown scaled down;
– Armaments Depot, Depot Support Group (Simonstown) scaled down;
– Naval Dockyard, Durban scaled down;
– Armaments Depot, Stores Depot (Durban) scaled down; and
– SAS *Wingfield*: Stores Depot, General Naval Workshop scaled down.

## IV. Medical service

The following rationalization and restructuring measures were undertaken:

– Quartermaster stores of the medical service in Pretoria, Cape Town and Bloemfontein consolidated into one Quartermaster organization;
– Medical Service training centre in Potchefstroom closed and moved to Pretoria;
– Medical Supply Depots 12, 13 and 15 closed; and
– sick bays and military medical clinics consolidated and scaled down according to the closing or scaling down of army, air force and navy units, bases and installations.

# Appendix 2. Armscor organizational structure, 1992–93

|  | 1992 | 1993 |
|---|---|---|
| **Board of Directors** | | |
| Chairman | I. Moolman | I. Moolman[a] |
| Managing Director | T. de Waal | T. de Waal[a, b] |
| Chief of the SADF | Gen. A. Liebenberg | Gen. G. Meiring[a] |
| SADF Chief of Staff Finance | Adm. A. Malherbe | Adm. A. Malherbe[a] |
| Managing Director: AEC | Dr D. Cronje | Dr D. Cronje |
| Director | L. Bartel | L. Bartel |
| Director | Prof. M. de Vries | Prof. M. de Vries |
| Director | P. van Rooy | P. van Rooy |
| Director | J. Alberts[c] | Adv. F. Bam[d] |
| **Management Board** | | |
| Managing Director | T. de Waal | T. de Waal[a] |
| Secretary | J. Kruger | J. Kruger |
| Public relations | J. Adler | A. Omar[d] |
| Finance | M. Coetzee | C. Hoffman |
| Quality control | J. Oosthuizen | J. Oosthuizen |
| Planning | Dr A. Buys | Dr A. Buys |
| Personnel | R. Petersen | J. Kgare[d] |
| Vehicles/Weapon systems | J. de Jager | J. de Jager |
| Electronics/Maritime | B. de Bruyn | B. de Bruyn |
| Aeronautics | H. de W. Esterhuyse | H. de W. Esterhuyse |
| Foreign procurement | R. Pretorius | R. Pretorius |
| Import/export control | P. Smith | P. Smith |

[a] Members of Defence Planning Committee (DPC).

[b] Former Director, Acquisition, Armscor; Denel Board of Directors, 1992.

[c] Managing Director, Denel.

[d] New appointments from outside Armscor.

*Source: Armscor Annual Report, 1992/93 and 1993/94.*

# Appendix 3. Denel organizational structure and diversification, 1992–93

|  | 1992 | 1993 |
|---|---|---|
| **Board of Directors** |  |  |
| Chairman | Dr J. Maree[a, b] | Dr J. Maree |
| Managing Director | J. Alberts[a, c] | J. Alberts |
| Director | T. de Waal[a, d] |  |
| Director | D. Brink | D. Brink |
| Director | D. Hawton | D. Hawton |
| Director | Dr J. Job | Dr J. Job |
| Director | M. Jonker | M. Jonker |
| Director | J. van Zyl | J. van Zyl |
| **Management Board** |  |  |
| Managing Director | J. Alberts[a] | J. Alberts |
| Aerospace | B. Kruger | B. Kruger |
| Systems | M. Koorts[a, e] | M. Koorts |
| Manufacturing | J. van Wyk[a, f] | T. Keuzenkamp[g] |
| Informatics | P. van den Heever[h] | P. van den Heever |
| Engineering | C. Veldman | C. Veldman[i] |
| Marketing | F. Stark | Dr F. Visser[j] |
| Finance | H. du Potgieter | H. du Potgieter |
| Planning | Dr T. van Reenen | Dr T. van Reenen |
| Personnel | A. Griesel | A. Griesel |
| Communications | P. Holtzhausen | P. Holtzhausen |
| Legal services | Dr J. Visser | F. Humphries |
| Promotions |  | F. Stark[k] |

[a] Former Armscor personnel.

[b] Former Executive Vice-Chairman of Armscor (1979–82); currently Chairman of Eskom and Nedcor; Director of Barlow Rand, Old Mutual and Development Bank of Southern Africa.

[c] Former Director, Finance and Planning, Armscor; Board of Directors, Armscor, 1992.

[d] Managing Director, Armscor.

[e] Former Executive Director, Vehicle and Weapons Industries, Armscor.

[f] Former Executive Director, Chemical Industries, Armscor.

[g] Former General Manager, Naschem, Denel.

[h] Former Director, Personnel, Armscor.

[i] Also Executive Director, Business Development Group, Denel.

[j] Former Director, Legal Services, Denel.

[k] Former Director, Marketing, Denel.

| 1992 | 1993 |
|---|---|
| **Corporate (Head Office)** | **Corporate** |
| Denel Insurance | Denel Insurance |
| Dinmar | Dinmar[a] |
| **Systems** | **Systems** |
| LIW[b] | LIW |
| Kentron[b] | Kentron |
| Eloptro[b] | Eloptro |
| Musgrave[b] | Musgrave |
| Mechem[b] | Mechem |
| **Manufacturing** | **Manufacturing** |
| Advena[b, c] | – |
| Naschem[b] | Naschem |
| PMP[b] | PMP |
| Somchem[b] | Somchem |
| Swartklip[b] | Swartklip |
| **Aerospace** | **Aerospace** |
| Houwteq[b] | Houwteq |
| Simera (Atlas)[b] | Simera |
| OTR[b] | OTR |
| **Engineering Services** | **Engineering** |
| DMS[d] | Dendex[e] |
| Gennan[b] | Gennan |
| (Gerotek)[b] | (Gerotek) |
| Mexa | Mexa |
| (Armatron)[b] | (Armatron) |
| (Ergotech) | (Ergotech) |
| **Informatics/Properties**[f] | **Informatics**[g] |
| Infoplan[b] | Infoplan |
| Excelsa | Excelsa |
| Denprop | Intersolve Health[h] |
| Mediamakers | |
| Bonaero Park | |
| | **Properties**[g] |
| | Denprop |
| | Bonaero Park |

[a] International Marketing Division.

[b] Former Armscor subsidiaries and business unit.

[c] Incorporated into PMP in 1993.

[d] Incorporated into Head Office in 1993.

[e] All the divisions of Engineering Services were rationalized into Dendex in 1993.

[f] The Informatics and Properties Group was divided into two separate groups in 1993.

[g] New industrial group in 1993.

[h] New business unit in 1993.

**Systems Group: military and civilian products**

| Division | Military | Civilian |
|---|---|---|
| LIW | Automatic cannon | Skid-steer loader[a] |
| | Small arms (pistols) | Ambidex tractor[b] |
| | Artillery systems | Mining equipment |
| | AFV guns/turrets | Machining |
| | Mortars | Drifter rock drill |
| Kentron[c] | Missiles | Turnkey security systems |
| | Air defence systems | Traffic engineering systems |
| | RPVs | Electric motors |
| | Weapon systems | 'Fibrestruc' glass fibre products |
| | Avionics | Solargen solar power system |
| | Inertial systems | Observation systems |
| | Sights | |
| | Observation systems | |
| | Fire-control systems | |
| | Stand-off weapons | |
| Eloptro | Electro-optical equipment | Acbit pattern display unit |
| | Sighting systems | Thermal imaging |
| | Laser rangefinders | Laser products |
| | Infra-red | Night vision products |
| Musgrave | Rifles | Sporting rifles |
| | Shotguns | Motor vehicle spares |
| | | Cricket bats |
| | | Wooden saw handles |
| | | Security products |
| Mechem | Mine-resistant vehicles[d] | Demolition |
| | Detonics | Bomb disposal |
| | Rocket systems | Mine clearance[e] |
| | Fuse design | Vapour detection |

*Notes*: AFV = armoured fighting vehicle; RPV = remotely piloted vehicle.

[a] Joint venture with Bell Pty (Ltd).

[b] In association with Dendex, Engineering Group.

[c] Kentron's commercial business is handled by Irenco, which was established by the Industrial Development Corporation (IDC) in 1990 and became part of Kentron in Oct. 1992.

[d] Licensing agreement with British company Alvis to produce mine-protected vehicles.

[e] Contract with United Nations for mine clearance in Mozambique.

**Systems Group financial results, 1992–93**

Figures are in m. rand in constant 1993 prices. Figures in italics are percentages.

| | 1992 | % of turnover | 1993 | % of turnover |
|---|---|---|---|---|
| Local defence | 598 | *60.6* | 512 | *52.6* |
| Exports | 326 | *33.0* | 390 | *40.0* |
| Commercial/other | 62 | *6.4* | 71 | *7.4* |
| **Total turnover** | **986** | *100.0* | **973** | *100.0* |

## Manufacturing Group: military and civilian products

| Division | Military | Civilian |
|---|---|---|
| Somchem | Artillery charges<br>Rocket systems<br>Anti-armour weapons<br>Missiles (air-to-air)<br>Propellants<br>Mortars<br>Automatic guns | Nitro-cellulose and chemical products[a]<br>Composite material products[b]<br>'Vectus' pipes[c]<br>Medical and electronic monitoring<br>  equipment<br>Food additives |
| Swartklip | Pyrotechnic products<br>Explosive devices<br>Artillery shells<br>Hand/rifle grenades<br>Electronic warfare equipment | Emergency flares<br>Sporting ammunition<br>Industrial power tools<br>Boulder Breaker system |
| Naschem[d] | Medium- and heavy-calibre<br>  ammunition | Mining explosives |
| PMP | Small and medium-calibre<br>  ammunition<br>Rapid-fire ammunition | Brass/copper products<br>Pressed/machined parts for motor<br>  industry<br>Pyrotechnic products<br>Explosive bonding, drill bits |

[a] For example, Nexus precision paints.

[b] Contract to supply composite material products for the SANAE IV base in Antarctica in 1993.

[c] Glass-reinforced polyester pipes manufactured under licence from Tubi Sarplast, Italy.

[d] Joint venture with Cementation Company (Africa) in La Forge (Pty) Ltd to produce grenades.

## Manufacturing Group financial results, 1992–93

Figures are in m. rand in constant 1993 prices. Figures in italics are percentages.

| | 1992 | % of turnover | 1993 | % of turnover |
|---|---|---|---|---|
| Local defence | 284 | *44.3* | 167 | *28.6* |
| Exports | 200 | *31.2* | 254 | *43.5* |
| Commercial/other | 157 | *24.5* | 162 | *27.9* |
| **Total turnover** | **641** | *100.0* | **583** | *100.0* |

## Aerospace Group: military and civilian products

| Division | Military | Civilian |
|---|---|---|
| Simera[a] | Design, development, manufacture, upgrading[b] and maintenance of fixed- and rotary-wing military aircraft[d] Trainer aircraft[e] Gas turbine engines Engine components | Design, development, manufacture, upgrading and maintenance of fixed- and rotary-wing commercial aircraft[c] Gas turbine engines Engine components Aircraft engine gearboxes[f] Airline maintenance |
| Houwteq[g] | | Low earth orbit (LEO) satellites Earth resource management system[h] |
| OTR | Test range for missiles, rockets, aircraft and satellites | Information processing and communication Launch site for LEOs |

[a] Simera incorporates Telcast, a foundry which produces aircraft engines, gearboxes and components using vacuum casting.

[b] Includes conversion of Aérospatiale SA-330 Puma helicopters to the Oryx configuration and the upgrade of Mirage R2Zs to the Cheetah E configuration.

[c] Passenger to freighter conversions of Airbus 300 B4 aircraft.

[d] Partnership with Marshall Engineering (UK) for CSH-2 Rooivalk attack helicopter.

[e] Composite material trainer (Ovid) developed in conjunction with Aerotek (CSIR).

[f] Contract with Rolls Royce (UK) to build gearboxes for RB211-535 aircraft engines.

[g] Houwteq was formerly involved in military satellites. It converted its production facilities to civilian purposes in 1992.

[h] In association with Alenia Spazio SpA, Italy.

## Aerospace Group financial results, 1992–93

Figures are in m. rand in constant 1993 prices. Figures in italics are percentages.

| | 1992 | % of turnover | 1993 | % of turnover |
|---|---|---|---|---|
| Local defence | 528 | *50.0* | 662 | *77.6* |
| Exports | 0 | .. | 25 | *2.9* |
| Commercial | 530 | *50.0* | 161 | *19.5* |
| **Total turnover** | **1 058** | *100.0* | **848** | *100.0* |

## Informatics and Properties groups: military and civilian products

| Division | Military | Civilian |
|---|---|---|
| **Informatics Group**[a] | | |
| Informatics[b] | Military computers and software[d] | Computers/software for health care[c] |
| | | Computers/software for manufacturing[e] |
| | | Networking[f] |
| | | Card technology[g] |
| **Properties Group** | | |
| Denprop | | Property development |
| Bonaero Park | | Property management |

[a] The Informatics and Properties Group was split into two separate groups in 1993. Each group was restructured in order to be more commercially oriented.

[b] The Informatics Group was made up of a number of business units each serving different markets: Infoplan (defence); Intersolve (health); Excelsa (manufacturing); Infovan (networking) and ID Technologies (card technology).

[c] Health informatics handled by Intersolve.

[d] Military informatics handled by Infoplan.

[e] Informatics for the manufacturing sector handled by Excelsa.

[f] Networking handled by Infovan.

[g] Card technology handled by ID Technologies.

## Informatics Group financial results, 1992–93

Figures are in m. rand in constant 1993 prices. Figures in italics are percentages.

| | 1992 | % of turnover | 1993 | % of turnover |
|---|---|---|---|---|
| Local defence | 265 | *85.0* | 278 | *80.0* |
| Commercial | 47 | *15.0* | 69 | *20.0* |
| **Total turnover** | **312** | ***100.0*** | **347** | ***100.0*** |

## Properties Group financial results, 1992–93

Figures are in m. rand in constant 1993 prices. Figures in italics are percentages.

| | 1992 | % of turnover | 1993 | % of turnover |
|---|---|---|---|---|
| Commercial | 45 | *100* | 66 | *100* |
| **Turnover** | **45** | ***100*** | **66** | ***100*** |

**Engineering Group: military and civilian products**

| Division | Military | Civilian |
|---|---|---|
| Dendex[a] | Systems analysis[b]<br>Military logistics<br>Systems engineering[c]<br>Test facility[d] | Systems analysis<br>Industrial ergonomics<br>Systems engineering<br>Operational support[e]<br>Ambidex tractor[f]<br>Product evaluation[g] |

[a] During 1993 the Engineering Group was restructured, and the former divisions (Mexa and Gennan) were rationalized into one division, Dendex. DMS became part of the Head Office in 1993. Dendex was made up of a number of business units: Gennan, Gerotek, Mexa, Ergotech and Industek. Dendex is also the site of Denel's business development unit, which is a multi-professional unit aimed at developing new commercial products.

[b] Military systems analysis carried out by Mexa.

[c] Military systems engineering carried out by Gennan.

[d] Test facility (Gerotek) includes environmental laboratory and vehicle test track.

[e] Dendex was awarded operational support contracts by Sasol and Columbus Project in 1993.

[f] Ambidex tractor developed by Dendex and manufactured by LIW (Systems Group).

[g] Contract for tyre tests for Continental Tyres (Germany) awarded to Gerotek in 1993.

**Engineering Group financial results, 1992–93**

Figures are in m. rand in constant 1993 prices. Figures in italics are percentages.

| | 1992 | % of turnover | 1993 | % of turnover |
|---|---|---|---|---|
| Local defence | 37 | *88.0* | 25 | *82.0* |
| Commercial | 5 | *12.0* | 5 | *18.0* |
| **Total turnover** | **42** | ***100.0*** | **30** | ***100.0*** |

*Sources for this appendix: Armscor Annual Report 1992/93* and *1993/93*; *Denel Annual Report 1992/93* and *1993/94*; *Engineering News*, 12 Mar. 1993; 'Armaments industry', *Business Day*, 7 Oct. 1992; and interviews with company officials.

# Appendix 4. Background statistics

**Military expenditure, the arms industry and the South African economy, 1961–89**

| | 1961 | 1965 | 1968 | 1970 | 1975 | 1977 | 1980 | 1985 | 1986 | 1987 | 1988 | 1989 |
|---|---|---|---|---|---|---|---|---|---|---|---|---|
| GDP deflator[a] | 11.5 | 12.7 | 14.3 | 15.8 | 27.5 | 33.7 | 53.3 | 100 | 115.4 | 130.8 | 151.4 | 174 |
| **Military expenditure** | | | | | | | | | | | | |
| Current m. rand | 71 | 229 | 254 | 257 | 970 | 1 654 | 1 890 | 4 274 | 5 123 | 6 683 | 8 195 | 9 937 |
| 1985 m. rand | 617 | 1 803 | 1 776 | 1 627 | 3 527 | 4 908 | 3 546 | 4 274 | 4 439 | 5 109 | 5 413 | 5 711 |
| **SADF procurement exp.** | | | | | | | | | | | | |
| Current m. rand | 8 | 66 | 127 | 117 | 626 | 1 047 | 944 | 1 904 | 2 327 | 3 489 | 3 825 | 5 816 |
| 1985 m. rand | 70 | 520 | 888 | 741 | 2 276 | 3 107 | 1 771 | 1 904 | 2 016 | 2 667 | 2 526 | 3 343 |
| As share of total military exp. (%) | 11.3 | 28.8 | 50.0 | 45.5 | 64.5 | 63.3 | 49.9 | 44.5 | 45.4 | 52.2 | 46.7 | 58.5 |
| **Military R&D expenditure** | | | | | | | | | | | | |
| Current m. rand | .. | .. | .. | .. | .. | 29 | 71 | 259 | 487 | 720 | 780 | 856 |
| 1985 m. rand | .. | .. | .. | .. | .. | 86 | 133 | 259 | 422 | 550 | 515 | 492 |
| R&D as share of total military exp. (%) | .. | .. | .. | .. | .. | 1.8 | 3.8 | 6.1 | 9.5 | 10.8 | 9.5 | 8.6 |
| **Total R&D expenditure** | | | | | | | | | | | | |
| Current m. rand | .. | .. | .. | .. | .. | .. | .. | 1 077 | .. | 1 329 | .. | 1 775 |
| 1985 m. rand | .. | .. | .. | .. | .. | .. | .. | 1 077 | .. | 1 016 | .. | 1 020 |
| Mil. R&D as share of total R&D | .. | .. | .. | .. | .. | .. | .. | 24.0 | .. | 54.2 | .. | 48.2 |
| **Domestic arms production (including exports)** | | | | | | | | | | | | |
| Current m. rand | 3 | 17 | 27 | 29 | 156 | 292 | 909 | 1 386 | 1 681 | 2 478 | 2 487 | 3 578 |
| 1985 m. rand | 26 | 134 | 189 | 184 | 567 | 866 | 1 705 | 1 386 | 1 457 | 1 894 | 1 643 | 2 056 |

| | 1961 | 1965 | 1968 | 1970 | 1975 | 1977 | 1980 | 1985 | 1986 | 1987 | 1988 | 1989 |
|---|---|---|---|---|---|---|---|---|---|---|---|---|
| GDP deflator | 11.5 | 12.7 | 14.3 | 15.8 | 27.5 | 33.7 | 53.3 | 100 | 115.4 | 130.8 | 151.4 | 174 |
| **Arms exports** | | | | | | | | | | | | |
| Current m. rand | .. | .. | .. | .. | .. | .. | .. | 282 | 331 | 454 | 269 | 205 |
| 1985 m. rand | .. | .. | .. | .. | .. | .. | .. | 282 | 287 | 347 | 178 | 118 |
| **Arms imports** | | | | | | | | | | | | |
| Current m. rand | .. | .. | .. | 95 | 413 | .. | 453 | 800 | 977 | 1 465 | 1 607 | 2 443 |
| Import prices (1985=100) | .. | .. | .. | 13.4 | 26.2 | .. | 53.5 | 100 | 122.6 | 134.5 | 149.3 | 173.6 |
| 1985 m. rand | .. | .. | .. | 709 | 1 576 | .. | 847 | 800 | 797 | 1 089 | 1 076 | 1 407 |
| **GDP** | | | | | | | | | | | | |
| Current m. rand | 5 406 | 7 682 | 10 125 | 12 473 | 26 646 | 33 263 | 60 328 | 123 126 | 142 135 | 164 524 | 200 448 | 240 639 |
| 1985 m. rand | 49 239 | 64 303 | 74 980 | 82 637 | 98 850 | 100 979 | 115 114 | 123 126 | 123 148 | 125 735 | 131 016 | 134 025 |
| Military expenditure as share of GDP (%) | 1.3 | 3.0 | 2.5 | 2.1 | 3.6 | 5.0 | 3.1 | 3.5 | 3.6 | 4.1 | 4.1 | 4.1 |
| **Total government exp.** | | | | | | | | | | | | |
| Current m. rand | 920 | 1 530 | 2 272 | 2 620 | 6 809 | 9 115 | 14 329 | 35 146 | 42 198 | 50 055 | 60 697 | 76 396 |
| 1985 m. rand | 8 000 | 12 047 | 15 888 | 16 582 | 24 760 | 27 047 | 26 884 | 35 146 | 36 567 | 38 268 | 40 090 | 43 906 |
| Military exp. as share of government exp. (%) | 7.7 | 15.0 | 11.2 | 9.8 | 14.2 | 18.1 | 13.2 | 12.2 | 12.1 | 13.4 | 13.5 | 13.0 |
| **Manufacturing output** | | | | | | | | | | | | |
| Current m. rand | 1 103 | 1 737 | 2 185 | 2 780 | 5 961 | 6 923 | 12 934 | 25 928 | 30 277 | 35 752 | 44 385 | 54 207 |
| 1985 m. rand | 9 591 | 13 677 | 15 280 | 17 595 | 21 676 | 20 543 | 24 266 | 25 928 | 26 237 | 27 333 | 29 316 | 31 153 |
| Arms as share of total manfct. output (%) | 0.3 | 1.0 | 1.2 | 1.0 | 2.6 | 4.2 | 7.0 | 5.3 | 5.6 | 6.9 | 5.6 | 6.6 |
| **Total exports** | | | | | | | | | | | | |
| Current m. rand | 1 691 | 2 071 | 2 652 | 2 747 | 7 480 | 10 339 | 22 022 | 39 973 | 45 856 | 48 791 | 56 923 | 66 317 |
| 1985 m. rand | 14 704 | 16 307 | 18 545 | 17 386 | 27 200 | 30 680 | 41 317 | 39 973 | 39 737 | 37 302 | 37 598 | 38 113 |
| Arms as % of total | .. | .. | .. | .. | .. | .. | .. | 0.7 | 0.7 | 0.9 | 0.5 | 0.3 |

| | | | | | | | | | | | | |
|---|---|---|---|---|---|---|---|---|---|---|---|---|
| **Manufactured exports** | | | | | | | | | | | | |
| Current m. rand | .. | .. | .. | 993 | 1 893 | 3 075 | 5 691 | 12 122 | 14 890 | 14 701 | 18 435 | 22 976 |
| 1985 m. rand | .. | .. | .. | 6 285 | 6 884 | 9 125 | 10 677 | 12 122 | 12 903 | 11 239 | 12 176 | 13 205 |
| Arms as share of manufact. exports (%) | .. | .. | .. | .. | .. | .. | .. | 2.3 | 2.2 | 3.1 | 1.5 | 0.9 |
| **Total imports** | | | | | | | | | | | | |
| Current m. rand | 1 198 | 2 136 | 2 324 | 3 189 | 8 128 | 8 491 | 17 034 | 28 409 | 31 981 | 35 273 | 46 544 | 53 950 |
| Import prices (1985=100) | 11.3 | 12.0 | 12.8 | 13.4 | 26.2 | 33.8 | 53.5 | 100 | 122.6 | 134.5 | 149.3 | 173.6 |
| 1985 m. rand | 10 602 | 17 800 | 18 156 | 23 799 | 31 023 | 25 121 | 31 839 | 28 409 | 26 086 | 26 225 | 31 175 | 31 077 |
| Arms as share of total imports (%) | .. | .. | .. | 3.0 | 5.1 | .. | 2.7 | 2.8 | 3.1 | 4.2 | 3.5 | 4.5 |
| **Manufactured imports** | | | | | | | | | | | | |
| Current m. rand | .. | .. | .. | 2 678 | 6 066 | 5 717 | 11 120 | 19 645 | 21 945 | 25 507 | 35 722 | 40 969 |
| 1985 m. rand | .. | .. | .. | 19 985 | 23 153 | 16 914 | 20 785 | 19 645 | 17 900 | 18 964 | 23 926 | 23 600 |
| Arms imports as share of manufact. imports (%) | .. | .. | .. | 3.5 | 6.8 | .. | 4.1 | 4.1 | 4.5 | 5.7 | 4.5 | 6.0 |
| Total employment (th.) | 2 643 | 2 934 | 3 407 | 3 741 | 4 577 | 4 664 | 4 893 | 5 246 | 5 295 | 5 372 | 5 600 | 5 650 |
| Arms industry employment (th.) | 0.83 | 9.49 | 14.60 | 15.72 | 36.95 | 52.95 | 122.80 | 116.55 | 125.95 | 138.05 | 154.65 | 131.75 |
| Arms industry as share of total employ't (%) | 0.0 | 0.3 | 0.4 | 0.4 | 0.8 | 1.1 | 2.5 | 2.2 | 2.4 | 2.6 | 2.8 | 2.3 |
| Manufacturing employment (th.) | 673 | 914 | 989 | 1 070 | 1 313 | 1 324 | 1 460 | 1 484 | 1 470 | 1 505 | 1 579 | 1 583 |
| Arms industry as share of manufacturing employment (%) | 0.1 | 1.0 | 1.5 | 1.5 | 2.8 | 4.0 | 8.4 | 7.9 | 8.6 | 9.2 | 9.8 | 8.3 |

*Note*: All current price series are deflated by the GDP deflator (1985=100) unless otherwise stated.

**Military expenditure, the arms industry and the South African economy, 1989–95**

| | 1989 | 1990 | 1991 | 1992 | 1993 | 1994 | 1995 |
|---|---|---|---|---|---|---|---|
| GDP deflator[a] | 86.9 | 100 | 113.5 | 127.6 | 141.7 | 155.2 | 168.6 |
| **Military expenditure** | | | | | | | |
| Current m. rand | 9 937 | 10 070 | 9 187 | 9 704 | 9 337 | 11 102 | 10 535 |
| 1990 m. rand | 11 435 | 10 070 | 8 094 | 7 605 | 6 589 | 7 153 | 6 249 |
| **Armscor procurement expenditure[b]** | | | | | | | |
| SANDF Special Defence Account (current m. rand) | .. | 5 414 | 4 598 | 3 678 | 3 898 | 3 112 | 2 999 |
| SANDF General Defence Account (current m. rand) | .. | .. | .. | 12 | 265 | 341 | 445 |
| SA Police Service (current m. rand) | .. | 90 | 171 | 210 | 274 | 271 | 175 |
| Other (current m. rand) | .. | .. | .. | 238 | 44 | 43 | 35 |
| Total (current m. rand) | .. | 5 504 | 4 769 | 4 138 | 4 481 | 3 767 | 3 654 |
| Total (1990 m. rand) | .. | 5 504 | 4 202 | 3 243 | 3 162 | 2 427 | 2 167 |
| **SANDF arms procurement expenditure** | | | | | | | |
| Procurement (current m. rand) | 5 817 | 5 747 | 4 174 | 4 383 | 3 740 | 3 094 | 3 514 |
| Procurement (1990 m. rand) | 6 693 | 5 747 | 3 677 | 3 435 | 2 639 | 1 993 | 2 084 |
| Procurement as share of total military expenditure (%) | 58.5 | 57.1 | 45.4 | 45.2 | 40.1 | 27.9 | 33.4 |
| **Military R&D expenditure** | | | | | | | |
| Current m. rand | 856 | 793 | 658 | 596 | 485 | 531 | 577 |
| 1990 m. rand | 985 | 793 | 580 | 467 | 342 | 342 | 342 |
| R&D as share of total military expenditure (%) | 8.6 | 7.9 | 7.2 | 6.1 | 5.2 | 4.8 | 5.5 |
| **Total R&D expenditure** | | | | | | | |
| Current m. rand | 1 775 | .. | 2 786 | .. | 2 594 | .. | .. |
| 1990 m. rand | 2 043 | .. | 2 455 | .. | 1 831 | .. | .. |
| Military R&D as share of total R&D (%) | 48.2 | .. | 23.6 | .. | 18.7 | .. | .. |
| **Domestic arms production (including exports)** | | | | | | | |
| Current m. rand | 3 578 | 3 685 | 4 646 | 3 929 | 4 605 | 4 102 | 4 083 |
| 1990 m. rand | 4 117 | 3 685 | 4 093 | 3 079 | 3 250 | 2 643 | 2 422 |

| | | | | | | | |
|---|---|---|---|---|---|---|---|
| Domestic procurement from national production (excluding imports) | | | | | | | |
| Current m. rand | 3 373 | 3 522 | 3 852 | 3 441 | 3 719 | 3 248 | 3 050 |
| 1990 m. rand | 3 881 | 3 522 | 3 394 | 2 697 | 2 625 | 2 093 | 1 809 |
| Arms exports | | | | | | | |
| Current m. rand | 205 | 163 | 794 | 488 | 886 | 854 | 1 033 |
| 1990 m. rand | 236 | 163 | 700 | 382 | 625 | 550 | 613 |
| Arms imports | | | | | | | |
| Current m. rand | 2 443 | 1 982 | 917 | 697 | 762 | 519 | 604 |
| Import prices (1990=100) | 90.8 | 100 | 108.3 | 112.8 | 118.4 | 123.3 | 135.5 |
| 1990 m. rand | 2 691 | 1 982 | 847 | 618 | 644 | 421 | 446 |
| GDP | | | | | | | |
| Current m. rand | 240 639 | 276 060 | 310 074 | 340 963 | 383 695 | 431 711 | 484 621 |
| 1990 m. rand | 276 940 | 276 060 | 273 249 | 267 257 | 270 702 | 278 148 | 287 506 |
| Military expenditure as share of GDP (%) | 4.1 | 3.6 | 3.0 | 2.8 | 2.4 | 2.6 | 2.2 |
| Total government expenditure | | | | | | | |
| Current m. rand | 76 396 | 81 380 | 93 615 | 115 278 | 137 050 | 137 156 | 154 903 |
| 1990 m. rand | 87 913 | 81 380 | 82 480 | 90 343 | 96 718 | 88 374 | 91 876 |
| Military expenditure as share of government exp. (%) | 13.0 | 12.4 | 9.8 | 8.4 | 6.8 | 8.1 | 6.8 |
| Manufacturing output | | | | | | | |
| Current m. rand | 54 207 | 63 064 | 69 151 | 74 295 | 81 167 | 90 177 | 104 474 |
| 1990 m. rand | 62 379 | 63 064 | 60 926 | 58 225 | 57 281 | 58 104 | 61 966 |
| Arms as share of total manufacturing output (%) | 6.6 | 5.8 | 6.7 | 5.3 | 5.7 | 4.5 | 3.9 |
| Total exports | | | | | | | |
| Current m. rand | 66 317 | 69 487 | 74 589 | 68 035 | 79 214 | 90 022 | 101 144 |
| 1990 m. rand | 76 314 | 69 487 | 65 717 | 53 319 | 55 903 | 58 004 | 59 991 |
| Arms as share of total exports (%) | 0.3 | 0.2 | 1.1 | 0.7 | 1.1 | 0.9 | 1.0 |
| Manufactured exports | | | | | | | |
| Current m. rand | 22 976 | 24 652 | 26 564 | 25 482 | 27 930 | 34 477 | 46 667 |
| 1990 m. rand | 26 440 | 24 652 | 23 404 | 19 970 | 19 711 | 22 214 | 27 679 |
| Arms as share of manufactured exports (%) | 0.9 | 0.7 | 3.0 | 1.9 | 3.2 | 2.5 | 2.2 |

| | 1989 | 1990 | 1991 | 1992 | 1993 | 1994 | 1995 |
|---|---|---|---|---|---|---|---|
| GDP deflator | 86.9 | 100 | 113.5 | 127.6 | 141.7 | 155.2 | 168.6 |
| Total imports | | | | | | | |
|   Current m. rand | 53 950 | 53 984 | 59 180 | 52 489 | 58 926 | 79 472 | 98 443 |
|   1990 m. rand | 59 416 | 53 984 | 54 645 | 46 533 | 49 769 | 64 454 | 72 652 |
|   Arms as share of total imports (%) | 4.5 | 3.7 | 1.5 | 1.3 | 1.3 | 0.7 | 0.6 |
| Manufactured imports | | | | | | | |
|   Current m. rand | 40 970 | 39 942 | 43 780 | 42 855 | 49 957 | 66 172 | 82 393 |
|   1990 m. rand | 45 121 | 39 942 | 40 424 | 37 992 | 42 194 | 53 668 | 60 807 |
|   Arms as share of manufactured imports (%) | 6.0 | 5.0 | 2.1 | 1.6 | 1.5 | 0.8 | 0.7 |
| Total employment (th.) | 5 650 | 5 633 | 5 538 | 5 425 | 5 312 | 5 278 | 5 318 |
| Arms industry employment (th.) | 131.75 | 118.15 | 106.94 | 83.36 | 74.98 | 74.63 | 76.25 |
| Arms industry employment as share of total employment (%) | 2.3 | 2.1 | 1.9 | 1.5 | 1.4 | 1.4 | 1.4 |
| Manufacturing employment (th.) | 1 583 | 1 582 | 1 547 | 1 504 | 1 477 | 1 481 | 1 493 |
| Arms industry as share of manufacturing employment (%) | 8.3 | 7.5 | 6.9 | 5.5 | 5.1 | 5.0 | 5.1 |

[a] All current price series are deflated by the GDP deflator (1990=100) unless otherwise stated.

[b] Includes all procurement expenditure for SANDF, the police, the Department of Correctional Services and other departments.

# Select bibliography

Abedian, I. and Standish, B. (eds), *Economic Growth In South Africa* (Oxford University Press: Cape Town, 1992).

Adam, H. and Moodley, K., *South Africa's Negotiated Revolution* (David Philip: Cape Town, 1992).

African National Congress, *The Reconstruction and Development Programme: A Policy Framework* (Umanyano Publications: Johannesburg, 1994).

Albright, D., 'The legacy of the South African nuclear weapons program', Paper presented at the Conference on a Nuclear Policy for a Democratic South Africa, Cape Town, 11–13 Feb. 1994.

Archer, S., 'Defence expenditure and arms procurement in South Africa', eds J. Cock and L. Nathan, *War and Society: The Militarisation of South Africa* (David Philip: Cape Town, 1989), pp. 244–59.

*Armscor Annual Report* (Armscor: Pretoria, various years).

Armscor, *The Armaments Industry in South Africa* (Armscor: Pretoria, 1984).

___, *South African Defence Industries Directory* (various years).

Atkinson, D., 'Brokering a miracle? The Multiparty Negotiating Forum', eds S. Friedman and D. Atkinson, *South African Review 7: The Small Miracle. South Africa's Negotiated Settlement* (Ravan Press: Johannesburg, 1994).

Baker, P., Boraine, A. and Krafchik, W. (eds), *South Africa and the World Economy in the 1990s* (David Philip: Cape Town, 1993).

Batchelor, P., 'Conversion of the South African defence industry: prospects and problems', Paper presented at workshop on Arms Trade and Arms Conversion in a Democratic South Africa, Military Research Group, Pretoria, 28–30 June 1993.

___, 'Disarmament, small arms and intra-state conflict: the case of Southern Africa', eds P. Batchelor, J. Potgieter and C. Smith, *Small Arms Management and Peace-keeping in Southern Africa* (United Nations Institute for Disarmament Research (UNIDIR): Geneva, 1996).

___, 'The economics of South Africa's arms trade', in *Budget Project Discussion Paper*, no. 3 (University of Cape Town, School of Economics: Cape Town, Aug. 1995).

___, 'Militarisation, disarmament and defence industrial adjustment: the case of South Africa', Unpublished PhD thesis, University of Cambridge (1996).

___, 'The origins and development of South Africa's defence industry', Paper presented at Arms Industry Workshop, Group for Environmental Monitoring and Centre for Conflict Resolution, Johannesburg, 14–15 Feb. 1996.

Baynham, S., 'After the cold war: political and security trends in Africa', *Africa Insight*, vol. 24, no. 1 (1994).

___, 'The new conditionality: swords into ploughshares?', *Africa Institute Bulletin*, vol. 31, no. 8 (1991).

___, 'Security strategies for a future South Africa', *Journal of Modern African Studies*, vol. 28, no. 3 (1990), pp. 401–30.

Baynham, S., 'Towards peace and security in Southern Africa', *Africa Institute Bulletin*, vol. 34, no. 3 (1993).

Benjamin, L., 'The Third World and its security dilemma', *International Affairs Bulletin*, vol. 14, no. 3 (1990).

Bethlehem, R., 'Issues of economic restructuring', ed. R. Schrire, *Wealth or Poverty? Critical Choices for South Africa* (Oxford University Press: Cape Town, 1992).

Bienen, H., 'Economic interests and security issues in Southern Africa', eds R. Rotberg *et al.*, *South Africa and its Neighbours: Regional Security and Self Interest* (Lexington Books: Lexington, Mass., 1985).

Birch, C. (ed.), *The New South Africa: Prospects for Security and Stability* (King's College, London, Centre for Defence Studies: London, 1994).

Bischoff, P., 'Democratic South Africa and the world order one year after: towards a new foreign policy script', *South African Perspectives*, no. 46 (1995).

Booth, K. and Smith, S. (eds), *International Relations Today* (Polity Press: Cambridge, 1995).

Booth, K. and Vale, P., 'Security in southern Africa after apartheid: beyond realism', *International Affairs*, vol. 71, no. 2 (1995).

Breytenbach, W., 'Conflict in Southern Africa: whither collective security?', *Africa Insight*, vol. 24, no. 1 (1994), pp. 26–37.

Brzoska, M., 'Arming South Africa in the shadow of the UN arms embargo', *Defence Analysis*, vol. 7, no. 1 (1991), pp. 21–38.

___, 'South Africa: evading the embargo', eds M. Brzoska and T. Ohlson, SIPRI, *Arms Production in the Third World* (Taylor & Francis: London, 1986).

Buys, A., 'The case for the retention of the arms industry', Paper presented at workshop on Arms Trade and Arms Conversion in a Democratic South Africa, Military Research Group, Pretoria, 28–30 June 1993.

___, 'The future of the South African armaments industry', *South African Defence Review*, no. 7 (1992), pp. 5–9.

Buzan, B., *People, States and Fear: An Agenda For International Security Studies in the Post Cold War Era* (Pinter: London, 1990).

Cameron Commission of Inquiry into Alleged Arms Transactions between Armscor and one Eli Wazan and other Related Matters, *First Report* (Cameron Commission: Johannesburg, June 1995).

___, *Second Report* (Cameron Commission: Cape Town, Nov. 1995).

Campbell, H., 'The dismantling of the apartheid military and the problems of conversion of the military industrial complex', Paper presented at CODESRIA Conference, Accra, 21–23 Apr. 1993.

Cawthra, G., *Brutal Force: The Apartheid War Machine* (International Defence and Aid Fund: London, 1986).

___, 'South Africa at war', ed. J. Lonsdale, *South Africa in Question* (James Currey: London, 1988).

Cilliers, J., 'The evolving security architecture in southern Africa', *African Security Review*, vol. 4, no. 5 (1995), pp. 30–47.

Cilliers, J., 'The future of the South African arms industry', ed. J. Garba, *Towards Sustainable Peace and Stability in Southern Africa* (Institute of International Education: New York, 1994).

___, 'To sell or die: the future of the South African defence industry', *ISSUP Bulletin*, no. 1 (1994).

___, 'Towards a South African conventional arms trade policy', *African Security Review*, vol. 4, no. 4 (1995), pp. 3–20.

Clapham, C., 'The African setting', ed. G. Mills, *From Pariah to Participant: South Africa's Evolving Foreign Relations 1990–1994* (South African Institute of International Affairs: Johannesburg, 1994).

___, 'The African state', Paper presented at the Conference of the Royal African Society on Sub-Saharan Africa, Cambridge, Apr. 1991.

Cobbett, W., 'Apartheid's army and arms embargo', eds J. Cock and L. Nathan, *War and Society: The Militarisation of South Africa* (David Philip: Cape Town, 1989), pp. 232–43.

Cochran, M., 'Post-modernism, ethics and international political theory', *Review of International Studies*, vol. 21, no. 3 (1995), pp. 237–50.

Cock, J., 'The cultural and social challenge of demilitarisation', *NOD and Conversion*, no. 37 (July 1996).

___, 'Introduction', eds J. Cock and L. Nathan, *War and Society: The Militarisation of South Africa* (David Philip: Cape Town, 1989), pp. 1–13.

___, 'Light weapons proliferation in southern Africa as a social issue', Paper presented at the Arms Industry Workshop, Group for Environmental Monitoring and Centre for Conflict Resolution, Johannesburg, 14–15 Feb. 1996.

___, *Redefining Security: The Military and the Ecology of Southern Africa*, GEM Discussion Document (Group for Environmental Monitoring: Bloenfontein, Mar. 1994).

___, 'Rocks, snakes and South Africa's arms industry', 1992. Unpublished.

___, 'South Africa's arms industry: strategic asset or moral cesspool', Paper presented at the workshop on Arms Trade and Arms Conversion in a Democratic South Africa, Military Research Group, Pretoria, 28–30 June 1993.

___, and Nathan, L. (eds), *War and Society: The Militarisation of South Africa* (David Philip: Cape Town, 1989).

Coker, C., *South Africa's Security Dilemmas* (Praeger: New York, 1987).

Cole, K. (ed.), *Sustainable Development for a Democratic South Africa* (Earthscan: London, 1994).

Davies, R., 'Approaches to regional integration in the southern African context', *Africa Insight*, vol. 24, no. 1 (1994).

*Denel Annual Report* (Denel: Pretoria,, various years).

Development Bank of Southern Africa, *South Africa's Nine Provinces: A Human Development Report* (Development Bank of Southern Africa: Halfway House, 1994).

de Villiers, D. and de Villiers, I., *P.W.* (Tafelberg: Cape Town, 1984).

de Waal, T., 'Commercialisation of the defence industry: issues faced in the procurement of arms', *South African Defence Review*, no. 11 (1993), pp. 7–10.

de Wet, G. *et al.*, 'The peace dividend in South Africa', eds N. Gleditsch *et al.*, *The Peace Dividend* (Elsevier: Amsterdam, 1996).

Ellis, G., *Third Force: What Evidence?*, South African Institute of Race Relations Regional Topic Paper, vol. 93, no. 1 (1993).

Ellis, S., 'Africa after the cold war: new patterns of government and politics', *Development and Change*, vol. 27, no. 1 (1996), pp. 1–28.

Evans, M., *The Frontline States, South Africa and Southern African Security: Military Prospects and Perspectives* (University of Zimbabwe: Harare, 1986).

Fanaroff, B., 'The arms industry: industrial relations and industrial policy', *South African Defence Review*, no. 11 (1993), pp. 11–15.

___, 'A trade unionist perspective on the future of the armaments industry in South Africa', *South African Defence Review*, no. 7 (1992), pp. 10–14.

Fine, B., 'Defence expenditure and the post-apartheid economy', Paper presented to the Conference on the Future of the Military and Defence in South Africa, Johannesburg, 1990.

Frankel, P., *Pretoria's Praetorians* (Cambridge University Press: Cambridge, 1984).

Friedman, S. (ed.), *The Long Journey: South Africa's Quest for a Negotiated Settlement* (Ravan Press: Johannesburg, 1993).

___, and Atkinson, D. (eds), *South African Review 7: The Small Miracle. South Africa's Negotiated Settlement* (Ravan Press: Johannesburg, 1994).

Fukuyama, F., 'The next South Africa', *South Africa International*, vol. 22, no. 2 (1991), pp. 77–81.

Fuller, G., 'The next ideology', *Foreign Policy*, no. 98 (spring 1995).

Gann, L. and Duignan, P., *Hope for South Africa?* (Hoover Institution Press: Stanford, Calif., 1991).

Garba, J. (ed.), *Towards Sustainable Peace and Stability in Southern Africa* (Institute for International Education: New York, 1994).

Gelb, S. (ed.), *South Africa's Economic Crisis* (David Philip: Cape Town, 1991).

Geldenhuys, D., *The Diplomacy of Isolation: South African Foreign Policy-Making* (Macmillan: Johannesburg, 1984).

___, *Isolated States: A Comparative Analysis* (Cambridge University Press: Cambridge, 1990).

George, P., 'The impact of South Africa's arms sales policy on regional military expenditure, development and security', ed. Swedish Ministry of Foreign Affairs, *Säkerhet och utveckling i Afrika* [Security and development in Africa] (Utrikesdepartementet: Stockholm, Nov. 1995), appendix, pp. 237–96 [appendix in English].

Geyer, H., 'Industrial development policy in South Africa: past, present and future', *World Development*, vol. 17 (1989), pp. 379–96.

Grundy, K., *The Militarisation of South African Politics* (I. B. Tauris: London, 1986).

___, *The Rise of the South African Security Establishment: An Essay on the Changing Locus of State Power*, South African Institute of International Affairs Bradlow Series, no. 1 (South African Institute of International Affairs: Braamfontein, 1983).

Gutteridge, W., 'South Africa's defence and security forces: the next decade', ed. J. Spence, *Change in South Africa* (Pinter: London, 1994).

Halliday, F., 'The end of the cold war and international relations: some analytical and theoretical conclusions', eds K. Booth and S. Smith, *International Relations Today* (Polity Press: Cambridge, 1995).

Hanlon, J., *Apartheid's Second Front: South Africa's War against its Neighbours* (Penguin: London, 1986).

Harbor, B., 'Lessons from the West European arms conversion experience', Paper presented at a workshop on Arms Trade and Arms Conversion in a Democratic South Africa, Military Research Group, Pretoria, 28–30 June 1993.

Hartley, K. and Sandler, T., *The Economics of Defence* (Cambridge University Press: Cambridge, 1995).

Heitman, H., 'Reshaping South Africa's armed forces', *Jane's Intelligence Review Special Report*, no. 3 (1994).

___, 'The South African defence industry: present and future prospects', *South African Defence Review*, no. 7 (1992), pp. 15–25.

___, '*South African War Machine* (Central News Agency: Johannesburg, 1985).

___, 'South Africa's defence industry: challenges and prospects', *South African Defence Review*, no. 2 (1992), pp. 26–34.

Hough, M., 'Disarmament and arms control with specific reference to the RSA', *South African Defence Review*, no. 9 (1993), pp. 17–23.

___, 'The UN arms embargo against South Africa: an assessment', *Strategic Review for Southern Africa*, vol. 10, no. 2 (1988), pp. 22–38.

___, and van der Merwe, M., *Selected Official South African Strategic Perspectives: 1976–1987*, ISSUP Ad Hoc Publication no. 25 (Institute for Strategic Studies, University of Pretoria: Pretoria, 1988).

___, and du Plessis, A., *Selected Official South African Strategic Perspectives: 1989–1992*, ISSUP Ad Hoc Publication no. 29 (Institute for Strategic Studies, University of Pretoria: Pretoria, 1992).

Howe, G. and le Roux, P. (eds), *Transforming the Economy: Policy Options for South Africa* (Indicator Project: Durban, 1992).

Howlett, D. and Simpson, J., 'Nuclearisation and denuclearisation in South Africa', *Survival*, vol. 35, no. 3 (1993), pp. 154–73.

International Defence and Aid Fund, *The Apartheid War Machine,* Fact Paper on Southern Africa no. 8 (International Defence and Aid Fund: London, 1980).

James, W. (ed.), *The State of Apartheid* (Lynne Rienner: Boulder, Colo., 1987).

Jaster, R., *The Defence of White Power: South African Foreign Policy Under Pressure* (St Martin's Press: New York, 1989).

Joffe, A., Kaplan, D., Kaplinsky, R. and Lewis, D., *A Framework for Industrial Revival in South Africa* (Industrial Strategy Project, Development Policy Research Unit, University of Cape Town: Cape Town, 1993).

___, *Improving Manufacturing Performance in South Africa* (University of Cape Town Press: Cape Town, 1995).

___, 'Manufacturing change', *Indicator South Africa*, vol. 11, no. 2 (1994), pp. 53–60.

Johnson, P. and Martin, D. (eds), *Frontline Southern Africa* (Ryan Publishing: Peterborough, 1989).

Kaplan, D., 'Ensuring technological advance in the South African manufacturing industry: some policy proposals', eds A. Joffe *et al.*, *Improving Manufacturing Performance in South Africa* (University of Cape Town Press: Cape Town, 1995).

Kaplinsky, R., 'South African industrial performance and structure in a comparative context', Paper presented at the Industrial Strategy Meeting, Johannesburg, 6–10 July 1992.

Keller, E. and Picard, L. (eds), *South Africa in Southern Africa: Domestic Change and International Conflict* (Lynne Rienner: Boulder, Colo., 1989).

Kennedy, P., *The Rise and Fall of Great Powers: Economic Change and Military Conflict from 1500 to 2000* (Random House: New York, 1987).

Khan, C., 'Thoughts on the future of the South African armaments industry', *South African Defence Review*, no. 7 (1992), pp. 26–28.

Kruys, J., 'The defence posture of the SADF in the nineties: some geo-strategic determining factors', Paper presented to Institute of Strategic Studies Conference, University of Pretoria, 25 June 1992.

Landgren, S., SIPRI, *Embargo Disimplemented* (Oxford University Press: Oxford, 1989).

Leonard, R., *South Africa at War* (Laurence Hill: Westport, Conn., 1983).

Levy, B., *How Can South African Manufacturing Efficiently Create Employment? An Analysis of the Impact of Trade and Industrial Policy,* Discussion Paper on Aspects of the Economy of South Africa no. 1 (World Bank, Southern African Department: Washington, DC, 1992).

Leysens, A., 'South Africa's military strategic link with Latin America: past developments and future prospects', *International Affairs Bulletin*, vol. 15, no. 3 (1991), pp. 23–47.

Lipton, M., *Capitalism and Apartheid: South Africa 1910–1986* (David Philip: Cape Town, 1986).

Lonsdale, J. (ed.), *South Africa in Question* (James Currey: London, 1988).

McCarthy, C., 'Apartheid ideology and economic development policy', eds N. Nattrass and E. Ardington, *The Political Economy of South Africa* (Oxford University Press: Cape Town, 1990).

___, 'Industrial development and distribution', ed. R. Schrire, *Wealth or Poverty? Critical Choices for South Africa* (Oxford University Press: Cape Town, 1992).

___, 'Structural development of South African manufacturing industry: a policy perspective', *South African Journal of Economics*, vol. 56, no. 1 (1988), pp. 1–23.

MacMillan, J. and Linklater, A. (eds), *Boundaries in Question: New Directions in International Relations* (Pinter: London, 1995).

McMillan, S., 'Economic growth and military spending in South Africa', *International Interactions*, vol. 18, no. 1 (1992), pp. 35–50.

McWilliams, J., *Armscor: South Africa's Arms Merchant* (Brasseys: London, 1989).

Martin, D. and Johnson, P., *Apartheid Terrorism: The Destabilisation Report* (James Currey: London, 1989).

___, *Destructive Engagement* (Zimbabwe Publishing House: Harare, 1986).

Matthews, R., 'The development of the South African military industrial complex', *Defence Analysis*, vol. 4, no. 1 (1988), pp. 7–24.

Maynes, C., 'The new pessimism', *Foreign Policy*, no. 100 (summer 1995), pp. 33–49.

Mearsheimer, J., 'Back to the future: instability after the cold war', *International Security*, vol. 15 (1990), pp. 5–56.

Meiring, G., 'Taking the South African Army into the future', *African Defence Review*, no. 14 (1994), pp. 1–6.

Mills, G. (ed.), *From Pariah to Participant: South Africa's Evolving Foreign Relations 1990–1994* (South African Institute of International Affairs: Johannesburg, 1994).

Mills, G., 'South Africa and Africa: regional integration and security co-operation', Paper presented at the AIC Conference on Defence, Midrand, 11–12 Oct. 1994.

Minty, A., 'South Africa's military build-up: the region at war', eds P. Johnson and D. Martin, *Frontline Southern Africa* (Ryan Publishing: Peterborough, 1989).

Moll, T., 'Did the apartheid economy fail?', *Journal of Southern African Studies*, vol. 17, no. 2 (1991), pp. 271–91.

___, 'From booster to brake? Apartheid and economic growth in comparative perspective', eds N. Nattrass and E. Ardington, *The Political Economy of South Africa* (Oxford University Press: Cape Town, 1990).

Möller, B., 'The concept of non-offensive defence: implications for developing countries with specific reference to southern Africa', Paper presented at Sir Pierre Van Ryneveld Air Power Conference, Pretoria, Aug. 1994.

___, *Non-Offensive Defence as a Strategy for Small States*, Working Paper no. 5 (Centre for Peace and Conflict Research: Copenhagen, 1994).

Murphy, C. and Tooze, R., *The New International Political Economy* (Lynne Rienner: Boulder, Colo., 1991).

Mutumi, T., 'South Africa and the nuclear Non-Proliferation Treaty', *African Security Review*, vol. 4, no. 2 (1995), pp. 46–51.

Nathan, L., 'Beyond arms and armed forces: a new approach to security', *South African Defence Review*, vol. 4 (1992), pp. 12–21.

___, *Changing the Guard: Armed Forces and Defence Policy in a Democratic South Africa* (Human Sciences Research Council: Pretoria, 1995).

___, 'Riding the tiger: the integration of armed forces and post-apartheid military', eds P. Vale and H. Orbon, *South Africa's Uncertain Transition: Reflections from Zimbabwe's Experience* (University of Zimbabwe: Harare, 1991).

___, 'Towards a post-apartheid threat analysis', *Strategic Review for Southern Africa*, vol. 15, no. 1 (1993), pp. 43–71.

___, 'Who guards the guardians? An agenda for civil–military relations and military professionalism in South Africa', *Strategic Review for Southern Africa*, vol. 17, no. 1 (1995), pp. 47–74.

___, *With Open Arms: Confidence and Security Building Measures in Southern Africa*, Disarmament Topical Papers no. 14 (United Nations: New York, 1993).

___, and Honwana, J., *After the Storm: Common Security and Conflict Resolution in Southern Africa*, Arusha Papers no. 3 (Centre for Southern African Studies, University of the Western Cape: Bellville, 1995).

___, and Phillips, M., 'Cross currents: security developments under F. W. de Klerk', eds G. Moss and I. Obery, *South African Review 6: From Red Friday to Codesa* (Ravan Press: Johannesburg, 1992).

Nattrass, J., *The South African Economy: Its Growth and Change* (Oxford University Press: Oxford, 1981).

Nattrass, N., 'Economic power and profits in post-war manufacturing', eds N. Nattrass and E. Ardington, *The Political Economy of South Africa* (Oxford University Press: Cape Town, 1990).

___, and Ardington, E. (eds), *The Political Economy of South Africa* (Oxford University Press: Cape Town, 1990).

Navias, M., 'Towards a new South African arms trade policy', *South African Defence Review*, no. 13 (1993), pp. 38–49.

Ohlson, T., 'South Africa: from apartheid to multi-party democracy', *SIPRI Yearbook 1995: Armaments, Disarmament and International Security* (Oxford University Press: Oxford, 1995), pp. 117–45.

Ohlson, T., Stedman, S. and Davies, R., *The New is Not Yet Born: Conflict Resolution in Southern Africa* (Brookings Institution: Washington, DC, 1994).

Oxfam, *Embracing the Future: Avoiding the Challenge of World Poverty. Oxfam's Response to the World Bank's 'Vision' for the Bretton Woods System* (Oxfam Policy Department: Oxford, 1994).

Philip, K., 'The private sector and the security establishment', eds J. Cock and L. Nathan, *War and Society: The Militarisation of South Africa* (David Philip: Cape Town, 1989), pp. 202–16.

Phillips, M., 'The nuts and bolts of military power: the structure of the SADF', eds J. Cock and L. Nathan, *War and Society: The Militarisation of South Africa* (David Philip: Cape Town, 1989), pp. 16–27.

Pretorius, R., 'The South African defence industry: defending apartheid', Unpublished honours thesis, University of the Witwatersrand: Johannesburg (1983).

Ratcliffe, S., 'Forced relations: the state, crisis and the rise of militarism in South Africa', Unpublished honours thesis, University of the Witwatersrand, Johannesburg (1983).

___, 'The shifting dominance: the foundations of the armaments industry during the 1960s', *Africa Perspective*, vol. 25 (1984), pp. 24–39.

Republic of South Africa, Armaments Development and Production Act no. 57 of 1968 (as amended), *Statutes of the Republic of South Africa* (Government Printer: Pretoria, 1968).

Republic of South Africa, Constitution of the Republic of South Africa, Act no. 200 of 1993, *Government Gazette*, no. 15466 (28 Jan. 1993).

Rogerson, C., 'Defending apartheid: Armscor and the geography of military production in South Africa', *GeoJournal*, vol. 22, no. 3 (1990), pp. 241–50.

___, 'The question of defence conversion in South Africa: issues from the international experience', *Africa Insight*, vol. 25, no. 3 (1995), pp. 154–62.

Roherty, J., *State Security in South Africa: Civil–Military Relations under P. W. Botha* (M. E. Sharpe: London, 1992).

Rotberg, R., 'Decision making and the military in South Africa', eds R. Rotberg *et al., South Africa and its Neighbours: Regional Security and Self Interest* (Lexington Books: Lexington, Mass., 1985).

Rotberg, R. *et al.*, *South Africa and its Neighbours: Regional Security and Self-Interest* (Lexington Books: Lexington, Mass., 1985).

Roux, A., 'Defence, human capital and economic development in South Africa', *South African Defence Review*, no. 19 (1994), pp. 14–22.

Saul, J. and Gelb, S., *The Crisis in South Africa: Class Defence, Class Revolution* (ZED Press, London, 1986).

Saurin, J., 'The end of international relations: the state and international theory in the age of globalisation', eds J. MacMillan and A. Linklater, *Boundaries in Question: New Directions in International Relations* (Pinter: London, 1995).

Schrire, R. (ed.), *Critical Choices for South Africa: An Agenda for the 1990s* (Oxford University Press: Cape Town, 1990).

___, *Malan to de Klerk: Leadership in the Apartheid State* (Hurst & Co.: London, 1994).

___, *Wealth or Poverty? Critical Choices for South Africa* (Oxford University Press: Cape Town, 1992).

Seegers, A., 'Apartheid's military: its origins and development', ed. W. James, *The State of Apartheid* (Lynne Rienner: Boulder, Colo., 1987).

___, 'Current trends in South Africa's security establishment', *Armed Forces and Society*, vol. 18, no. 2 (1992), pp. 159–74.

___, *The Military in the Making of Modern South Africa* (I. B. Tauris: London, 1996).

___, 'The military in South Africa: a comparison and a critique', *South Africa International*, vol. 16, no. 4 (1986).

___, 'The South African security establishment: theoretical and comparative impressions', ed. M. Swilling, *Views on the South African State* (Human Sciences Research Council: Pretoria, 1990).

___, 'South Africa's national security management system 1972–1990', *Journal of Modern African Studies*, vol. 29, no. 2 (1991), pp. 253–73.

Shaw, M., 'Biting the bullet: negotiating democracy's defence', eds S. Friedman and D. Atkinson, *South African Review 7: The Small Miracle. South Africa's Negotiated Settlement* (Ravan Press: Johannesburg, 1994), pp. 228–56.

Shaw, T. and Leppan, E., 'South Africa: white power and the regional military–industrial complex', eds O. Aluko and T. Shaw, *Southern Africa in the 1980s* (George Allen & Unwin: London, 1985).

Shearer, J., 'Denuclearisation of Africa: the South African dimension', *Disarmament*, vol. 16, no. 2 (1993), pp. 171–85.

Siko, M. and Cawthra, G., 'South Africa: prospects for NOD in the context of collective regional security', *NOD and Conversion*, no. 33 (July 1995).

Simkins, C., 'The South African economy: problems and prospects', ed. J. Spence, *Change in South Africa* (Pinter: London, 1994).

Simpson, G., 'The politics and economics of the armaments industry in South Africa', eds J. Cock and L. Nathan, *War and Society: The Militarisation of South Africa* (David Philip: Cape Town, 1989), pp. 217–31.

Singh, R. P. and Wezeman, P., 'South Africa's arms production and exports', *SIPRI Yearbook 1995: Armaments, Disarmament and International Security* (Oxford University Press: Oxford, 1995), appendix 14E, pp. 569–82.

Smith, S., 'Mature anarchy, strong states and security', *Arms Control*, vol. 12 (1991).

___, 'The self-images of a discipline: a genealogy of international relations theory', eds K. Booth and S. Smith, *International Relations Today* (Polity Press: Cambridge, 1995).

Sole, D., 'Southern African foreign policy assumptions and objectives from Hertzog to De Klerk', *South African Journal of International Affairs*, vol. 2, no 1 (summer 1994), pp. 104–13.

South African Defence Industry Association, *Some Statistics of the SADIA Members of the Defence Industry* (SADIA: Johannesburg, Jan. 1996).

South African Department of Arts, Culture, Science and Technology, *South Africa's Green Paper on Science and Technology; Preparing for the 21st Century* (Government Printer: Pretoria, 1996).

South African Department of Defence, *White Paper on Defence and Armaments Production 1965–1967* (Government Printer: Pretoria, 1967).

___, *White Paper on Defence and Armaments Production 1969* (Government Printer: Pretoria, 1969).

___, *White Paper on Defence 1977* (Government Printer: Pretoria, 1977).

___, *White Paper on Defence and Armaments Supply 1979* (Government Printer: Pretoria, 1979).

___, *White Paper on Defence and Armaments Supply 1982* (Government Printer: Pretoria, 1982).

___, *White Paper on Defence and Armaments Supply 1984* (Government Printer: Pretoria, 1984).

___, *White Paper on Defence and Armaments Supply 1986* (Government Printer: Pretoria, 1986).

___, *Briefing on the Organisation and Functions of the South African Defence Force and the Armaments Corporation of South Africa Ltd* (Government Printer: Pretoria, 1987).

___, *Defence in a Democracy: White Paper on National Defence* (Government Printer: Pretoria, 1996).

___, *The National Defence Force in Transition: SANDF Annual Report 1994/95* (Government Printer: Pretoria, 1995).

South African Institute of Race Relations, *Race Relations Survey* (South African Institute for Race Relations: Johannesburg, various years).

South African Reserve Bank, *Quarterly Bulletin* (South African Reserve Bank: Pretoria, various years).

Southern African Development Community, A Framework and Strategy for Building the Community, SADC, Harare, Jan. 1993.

___, The SADC Organ on Politics, Defence and Security. Statement issued by the Meeting of SADC Ministers Responsible for Foreign Affairs, Defence and SADC Affairs, Gaborone, 18 Jan. 1996.

Southern African Development Community, Towards the Southern African Development Community: A Declaration by the Heads of State or Government of Southern African States, Windhoek, 1992.

Sparks, D., 'The peace dividend: Southern Africa after apartheid', *Indicator South Africa*, vol. 10, no. 2 (1993), pp. 27–32.

Spence, J. (ed.), *Change in South Africa* (Pinter: London, 1994).

Stadler, A., *The Political Economy of Modern South Africa* (St Martin's Press: New York, 1987).

Steyn, P., 'Aligning the South African defence industry with regional integration and security', *African Security Review*, vol. 4, no. 2 (1995), pp. 15–23.

___, 'Challenges and prospects for the SA defence industry: equipping the armed forces for the future', *South African Defence Review*, no. 11 (1993), pp. 1–6.

Storey, P., 'Moral and ethical issues raised by the defence industry', Paper presented at the Defence Industry Conference, Midrand, 11–12 Oct.1995.

Strange, S., 'An eclectic approach', eds C. Murphy and R. Tooze, *The New International Political Economy* (Lynne Rienner: Boulder, Colo., 1991).

Suttner, R., 'Some problematic questions in developing foreign policy after April 27 1994', *Southern African Perspectives*, no. 44 (1994).

Swilling, M. and Phillips, M., 'State power in the 1980s: from total strategy to counter-revolutionary warfare', eds J. Cock and L. Nathan, *War and Society: The Militarisation of South Africa* (David Philip: Cape Town, 1989), pp. 134–48.

Taylor, R., 'The end of apartheid as part of the end of history?', *South African Journal of International Affairs*, vol. 3. no. 1 (1995), pp. 22–32.

Terreblanche, S. and Nattrass, N., 'A periodisation of the political economy from 1910', eds N. Nattrass and E. Ardington, *The Political Economy of South Africa* (Oxford University Press: Cape Town, 1990).

Tickner, A., 'Re-visioning security', eds K. Booth and S. Smith, *International Relations Today* (Polity Press: Cambridge, 1995).

*Top Companies* (Financial Mail: Johannesburg, various years).

Transitional Executive Council Sub-Council on Defence, National Policy for the Defence Industry (TEC Sub-Council on Defence: Pretoria, Apr. 1994).

van Aardt, M., 'In search of a more adequate concept of security for South Africa', *South African Journal of International Affairs*, vol. 1, no. 1 (spring 1993), pp. 82–101.

van Nieuwkerk, A., 'Where is the voice of the people? Public opinion and foreign policy in South Africa', *South African Journal of International Affairs*, vol. 1, no. 2 (autumn 1994).

Väyrynen, R., 'The role of transnational corporations in the military sector of South Africa', *Journal of Southern African Affairs*, vol. 5, no. 2 (1980), pp. 199–255.

Willett, S., 'The legacy of a pariah state: South Africa's arms trade in the 1990s', *Review of African Political Economy*, no. 64 (1995), pp. 151–66.

___, *Open Arms for the Prodigal Son: The Future of South Africa's Arms Trade Policies*, BASIC Report no. 2 (British American Security Information Council: London, 1994).

Willett, S., 'Ostriches, wise old elephants and economic reconstruction in Mozambique', *International Peacekeeping*, vol. 2, no. 1 (spring 1995), pp. 34–55.

___, 'Policies for adjustment to lower military spending', Paper presented at a workshop on Arms Trade and Arms Conversion in a Democratic South Africa, Military Research Group, Pretoria, 28–30 June 1993.

___, and Batchelor, P., 'The South African defence dudget', Military Research Group, Johannesburg, 1994. Unpublished.

___, and Batchelor, P., *To Trade or Not to Trade: The Costs and Benefits of South Africa's Arms Trade*, Working Paper no. 9 (Military Research Group: Johannesburg, 1994).

Williams, R., *The Changing Parameters of South African Civil–Military Relations: Past, Present and Future Scenarios*, Working Paper no. 6 (Military Research Group: Johannesburg, 1993).

___, 'Non-offensive defence and South Africa: considerations on a post-modern military, mission redefinition and defensive restructuring', *NOD and Conversion*, no. 37 (July 1996).

___, *South Africa's New Defence Force: Progress and Prospects*, CSIS Africa Notes no. 170 (Centre for Strategic and International Studies: Washington, DC, Mar. 1995).

Wiseman, J., 'Militarism, militarisation and praetorianism in South Africa', *Africa*, vol. 58, no. 2 (1988).

# Index

development in 20
disarmament in 20, 51
military expenditure 20, 21, 68, 123
peacekeeping 123, 124–25
peace promotion 20
political pluralism and 49–50
political reforms in 50
regional conflicts 25, 32, 50, 118
security cooperation 52–53, 116,
  120–23
South Africa dominance 122, 123, 130
superpowers and 20, 49
Union of Soviet Socialist Republics
  and 53
see also under names of countries
Southern African Development
  Community (SADC) 21, 52, 53, 120,
  121, 122, 124:
    'A Framework and Strategy for
      Building the Community' 52, 121
    Organ on Politics, Defence and
      Security 121
Southern African Development Co-
  ordination Conference (SADCC) 52
sovereign state, concept of 21–22
Soweto uprisings 32
Space Affairs Act (1993) 72
space-launch vehicles 127
Special Defence Account 68, 70:
    figures 69, 220
    military expenditure on R&D 46, 75,
      155
    technology and 88, 89
Special Forces Regiment 133
Spesaero 106, 108
Spescom 103
state:
    civilian control of re-established 55
    military and 25
state of emergency lifted 54
State President's Office 55, 56
State Security Council (SSC) 55, 56
State Tender Board 29
Steyn, Lieutenant-General P. 57, 129, 146
strategic doctrine:
    arms exports and 138
    changes in 114, 131–35
    non-offensive 132, 134–37, 138,
      139–40
    offensive 132, 133, 134, 137
strategic environment, changes in 49–53,
  117

Sub-Council on Defence see TEC
submarines 131, 137, 138, 151, 152,
  153
Sub-Saharan Africa: arms exports to 17,
  204
Sudan 149
Swartklip 30, 37, 91, 94, 178, 213
Swaziland 50
Switzerland: arms exports 58, 70, 76, 151
Synertech 107
Syria 149, 205

Tafelberg, SAS 87
Taiwan 126, 203
tanks 131, 138, 139
Tanzania: military expenditure 123
taxes 176, 185, 202
TBVC armies 60, 61
technology, state investment and 26
TEC (Transitional Executive Council) 54,
  55:
    Sub-Council on Defence 59, 60, 61,
      144, 172
TEC (Transitional Executive Council)
  Act (1993) 55
TEC (Transitional Executive Council)
  Bill 55, 61
Technology Development Programme
  88
Technology Secretariat 46
Teklogic 107, 162
Telcast 37
Telkom 65
Telsis 107
TFM 162
Third Force 57
threat perceptions 51
threats, internal 117
Tongaat 39
Total Strategy:
    failure of 49, 54
    first articulation 1, 24
Trade and Industry Department 204
trade unions:
    arms industry and 146, 201
    conversion and 179, 181
Transitional Executive Council see TEC
Transkei 60
transnational companies 35–36, 183
Transnet 65
Truckmakers 103, 104, 107
Tubi Sarplast 94

# DATE DUE

| | | | |
|---|---|---|---|
| | | | |
| | | | |
| | | | |
| | | | |
| | | | |
| | | | |
| | | | |
| | | | |
| | | | |
| | | | |
| | | | |
| | | | |
| | | | |
| | | | |
| | | | |
| | | | |
| | | | |
| | | | |
| | | | |
| | | | |

HIGHSMITH #45230

Printed
in USA